Grand Challenges in Environmental Sciences

Committee on Grand Challenges in Environmental Sciences

Oversight Commission for the Committee on Grand Challenges in Environmental Sciences

National Research Council

NATIONAL ACADEMY PRESS
Washington, D.C.

NATIONAL ACADEMY PRESS • 2101 Constitution Ave., N.W. • Washington, DC 20418

NOTICE: The project that is the subject of this report was approved by the Governing Board of the National Research Council, whose members are drawn from the councils of the National Academy of Sciences, the National Academy of Engineering, and the Institute of Medicine. The members of the committee responsible for the report were chosen for their special competences and with regard for appropriate balance.

This project was supported by Grant No. DBI-9806286 between the National Academy of Sciences and the National Science Foundation. Any opinions, findings, conclusions, or recommendations expressed in this publication are those of the author(s) and do not necessarily reflect the view of the organizations or agencies that provided support for this project.

International Standard Book Number 0-309-07254-9
Library of Congress Control Number: 2001089192

Additional copies of this report are available from:

National Academy Press
2101 Constitution Ave., NW
Box 285
Washington, DC 20055

800-624-6242
202-334-3313 (in the Washington metropolitan area)
http://www.nap.edu

THE NATIONAL ACADEMIES

National Academy of Sciences
National Academy of Engineering
Institute of Medicine
National Research Council

The **National Academy of Sciences** is a private, nonprofit, self-perpetuating society of distinguished scholars engaged in scientific and engineering research, dedicated to the furtherance of science and technology and to their use for the general welfare. Upon the authority of the charter granted to it by the Congress in 1863, the Academy has a mandate that requires it to advise the federal government on scientific and technical matters. Dr. Bruce M. Alberts is president of the National Academy of Sciences.

The **National Academy of Engineering** was established in 1964, under the charter of the National Academy of Sciences, as a parallel organization of outstanding engineers. It is autonomous in its administration and in the selection of its members, sharing with the National Academy of Sciences the responsibility for advising the federal government. The National Academy of Engineering also sponsors engineering programs aimed at meeting national needs, encourages education and research, and recognizes the superior achievements of engineers. Dr. William A. Wulf is president of the National Academy of Engineering.

The **Institute of Medicine** was established in 1970 by the National Academy of Sciences to secure the services of eminent members of appropriate professions in the examination of policy matters pertaining to the health of the public. The Institute acts under the responsibility given to the National Academy of Sciences by its congressional charter to be an adviser to the federal government and, upon its own initiative, to identify issues of medical care, research, and education. Dr. Kenneth I. Shine is president of the Institute of Medicine.

The **National Research Council** was organized by the National Academy of Sciences in 1916 to associate the broad community of science and technology with the Academy's purposes of furthering knowledge and advising the federal government. Functioning in accordance with general policies determined by the Academy, the Council has become the principal operating agency of both the National Academy of Sciences and the National Academy of Engineering in providing services to the government, the public, and the scientific and engineering communities. The Council is administered jointly by both Academies and the Institute of Medicine. Dr. Bruce M. Alberts and Dr. William A. Wulf are chairman and vice chairman, respectively, of the National Research Council.

Preface

The relationship of our technological society with the environment has become a central topic of discussion in the academic world, in society at large, and even in U.S. presidential election campaigns. It is clear that the topic is of great importance; it is equally clear that the topic is complex and multifaceted, and has many aspects for which our current understanding is inadequate.

The National Science Foundation (NSF), long a supporter of research in the disciplinary sciences, has become a major supporter of the inherently interdisciplinary environmental sciences as well. In carrying out this role, NSF asked the National Research Council (NRC) to address the following question: "Of the many scientific challenges related to the environment, which few offer the greatest potential for investment; that is, what are the 'grand challenges' in environmental science?" In response, the NRC established the Committee on Grand Challenges in Environmental Sciences, comprising 17 members with a broad range of disciplinary backgrounds.

The committee met five times between January and October 1999 to solicit input, select the most compelling challenges, and formulate its recommendations. The diversity of specialization and expertise needed for this task was far greater than is the case for most NRC studies, and it is to the committee's credit that it was able to reach agreement on a small set of grand challenges and an even smaller set of immediate research investments. This volume presents the results of the committee's efforts.

The environmental sciences are not the sole purview of NSF. We anticipate that this report will be of interest as well to the National Institutes of Health, the Environmental Protection Agency, the Department of Energy, and other organizations both within and outside of government. I hope that all of these organizations will coordinate their support for the environmental sciences so the maximum progress will be achieved.

This report has been reviewed in draft form by individuals chosen for their technical expertise and diverse perspectives in accordance with procedures approved by the NRC's Report Review Committee for reviewing NRC and Institute of Medicine reports. The purpose of that independent review was to provide candid and critical comments to assist the NRC in making the published report as sound as possible and to ensure that the report meets institutional standards for objectivity, evidence, and responsiveness to the study charge. The review comments and draft manuscript remain confidential to protect the integrity of the deliberative process. We wish to thank the following individuals for their participation in the review of this report: Cynthia Beall, Case Western Reserve University; Kenneth Brink, Woods Hole Oceanographic Institution; Ingrid Burke, Colorado State University; Robert Dickinson, Georgia Institute of Technology; Thomas Dietz, George Mason University; John Dowling, Harvard University; Theodore Geballe, Stanford University; Jack Gibbons, National Academy of Engineering; Bernard Goldstein, Rutgers University; William Graf, Arizona State University; Geoffrey Heal, Columbia University; George M. Hornberger, University of Virginia; Raymond Jeanloz, University of California, Berkeley; Pamela Matson, Stanford University; Judith E. McDowell, Woods Hole Oceanographic Institution; Hugh Morris, El Dorado Gold Corporation; Gilbert Omenn, University of Michigan; Gordon Orians, University of Washington; David Pilbeam, Harvard University; Henry Vaux, University of California; Thomas J. Wilbanks, Oak Ridge National Laboratory; and Eric Wood, Princeton University.

The individuals listed above have provided many constructive comments and suggestions. It must be emphasized, however, that responsibility for the final content of this report rests entirely with the authoring committee and the NRC.

The staff of the NRC has been particularly helpful in the deliberations of the committee and the preparation of this report. Leah Probst's efforts in shepherding the multiple drafts and comments and in preparing the results for publication were vital. Laurie Geller and Paul Stern provided both intellectual and organizational contributions of great merit. The committee extends its highest commendation to David Policansky, whose knowledge of the environmental sciences and of the NRC and whose interpersonal abilities and diligence are exceptional and deeply appreciated.

> Thomas E. Graedel, Ph.D.
> Chair, Committee on Grand Challenges
> in the Environmental Sciences

Contents

Executive Summary

Scientists have long worked to understand the environment and humanity's place in it. The search for this knowledge grows in importance as rapid increases in human populations and economic development intensify the stresses human beings place on the biosphere and ecosystems. People want to be warned of major environmental changes and, if the environment is under threat, want to know how to respond. Fortunately, rapid increases in scientific capability—such as recent advances in computing power and molecular biology and new techniques for sensing biological, physical, and chemical phenomena below, on, and above the Earth's surface—together with the rediscovery that the human-environment relationship is a critical topic for the human sciences, are making it possible for science to provide much of this knowledge. The scientific excitement and challenge of understanding the complex environmental systems humans depend on make the environmental sciences centrally important as humankind attempts a transition to a more sustainable relationship with the Earth and its natural resources.

This report was written in response to a request from the National Science Foundation (NSF) that the National Research Council (NRC), drawing on expertise from across the environmental sciences, offer a judgment regarding the most important environmental research challenges of the next generation—the areas most likely to yield results of major scientific and practical importance if pursued vigorously now. In formulating this judgment, the committee established by the NRC confronted the problem of the unity of the environment—the fact that every aspect of the environment is connected to every other in some way.

Consequently, no branch of environmental science can progress very far without drawing on knowledge from other branches.

The committee sought to identify a small number of grand challenges in the environmental sciences—major scientific tasks that are compelling for both intellectual and practical reasons, that offer potential for major breakthroughs on the basis of recent developments in science and technology, and that are feasible given current capabilities and a serious infusion of resources. After soliciting input as broadly as possible and considering more than 200 nominations from the scientific community, the committee selected the eight grand challenges described below. The committee's selection criteria included probability of significant scientific and practical payoff, large scope, relevance to important environmental issues, feasibility, timeliness, and requirement for multidisciplinary collaboration.

Attaining the needed environmental knowledge for the next generation will depend on the active pursuit of all eight grand challenges. However, the committee was asked to identify an even more focused list of activities to be pursued in the near term by NSF, either alone or in collaboration with other research funders. Therefore, the committee selected four areas, derived from the grand challenges, to recommend for immediate research investment by NSF and others. In addition to the criteria used to choose the eight grand challenges, the committee considered whether the activities are currently underfunded, i.e., stand to benefit from an infusion of financial and human resources; the committee also applied the criteria of scientific importance, urgency, and scope. The committee did not rank-order the grand challenges, as we consider them all to be broadly and deeply important, nor did we rank-order the immediate research investments for the same reason. Both are therefore presented below in alphabetical order.

THE GRAND CHALLENGES

1. Biogeochemical Cycles

The challenge is to further our understanding of the Earth's major biogeochemical cycles, evaluate how they are being perturbed by human activities, and determine how they might better be stabilized. Important research areas include quantifying the sources and sinks of the nutrient elements and gaining a better understanding of the biological, chemical, and physical factors regulating transformations among them; improving understanding of the interactions among the various biogeochemical cycles; assessing anthropogenic perturbations of biogeochemical cycles and their impacts on ecosystem functioning, atmospheric chemistry, and human activities, and developing a scientific basis for societal decisions about managing these cycles; and exploring technical and institutional approaches to managing anthropogenic perturbations.

2. Biological Diversity and Ecosystem Functioning

The challenge is to improve understanding of the factors affecting biological diversity and ecosystem structure and functioning, including the role of human activity. Important research areas include improving tools for rapid assessment of diversity at all scales; producing a quantitative, process-based theory of biological diversity at the largest possible variety of spatial and temporal scales; elucidating the relationship between diversity and ecosystem functioning; and developing and testing techniques for modifying, creating, and managing habitats that can sustain biological diversity, as well as people and their activities.

3. Climate Variability

The challenge is to increase our ability to predict climate variations, from extreme events to decadal time scales; to understand how this variability may change in the future; and to assess realistically the resulting impacts. Important research areas include improving observational capability, extending the record of observations back into the Earth's history, improving diagnostic process studies, developing increasingly comprehensive models, and conducting integrated impact assessments that take human responses and impacts into account.

4. Hydrologic Forecasting

The challenge is to develop an improved understanding of and ability to predict changes in freshwater resources and the environment caused by floods, droughts, sedimentation, and contamination. Important research areas include improving understanding of hydrologic responses to precipitation, surface water generation and transport, environmental stresses on aquatic ecosystems, the relationships between landscape changes and sediment fluxes, and subsurface transport, as well as mapping groundwater recharge and discharge vulnerability.

5. Infectious Disease and the Environment

The challenge is to understand ecological and evolutionary aspects of infectious diseases; develop an understanding of the interactions among pathogens, hosts/receptors, and the environment; and thus make it possible to prevent changes in the infectivity and virulence of organisms that threaten plant, animal, and human health at the population level. Important research areas include examining the effects of environmental changes as selection agents on pathogen virulence and host resistance; exploring the impacts of environmental change on disease etiology, vectors, and toxic organisms; developing new approaches to surveillance and monitoring; and improving theoretical models of host-pathogen ecology.

6. Institutions and Resource Use

The challenge is to understand how human use of natural resources is shaped by institutions such as markets, governments, international treaties, and formal and informal sets of rules that are established to govern resource extraction, waste disposal, and other environmentally important activities. Important research areas include documenting the institutions governing critical lands, resources, and environments; identifying the performance attributes of the full range of institutions governing resources and environments worldwide, from local to global levels; improving understanding of change in resource institutions; and conceptualizing and assessing the effects of institutions for managing global commons.

7. Land-Use Dynamics

The challenge is to develop a systematic understanding of changes in land uses and land covers that are critical to ecosystem functioning and services and human welfare. Important areas for research include developing long-term, regional databases for land uses, land covers, and related social information; developing spatially explicit and multisectoral land-change theory; linking land-change theory to space-based imagery; and developing innovative applications of dynamic spatial simulation techniques.

8. Reinventing the Use of Materials

The challenge is to develop a quantitative understanding of the global budgets and cycles of materials widely used by humanity and of how the life cycles of these materials (their history from the raw-material stage through recycling or disposal) may be modified. Important research areas include developing spatially explicit budgets for selected key materials; developing methods for more complete cycling of technological materials; determining how best to utilize materials that have uniquely useful industrial applications but are potentially hazardous to the environment; developing an understanding of the patterns and driving forces of human consumption of resources; and developing models for possible global scenarios of future industrial development and associated environmental implications.

RECOMMENDED IMMEDIATE RESEARCH INVESTMENTS

The committee recommends that immediate investments be made in four priority research areas related to the grand challenges.

1. Biological Diversity and Ecosystem Functioning

Recommendation: Develop a comprehensive understanding of the relationship between ecosystem structure and functioning and biological diversity. This initiative would include experiments, observations, and theory, and should have two interrelated foci: (a) developing the scientific knowledge needed to enable the design and management of habitats that can support both human uses and native biota; and (b) developing a detailed understanding of the effects of habitat alteration and loss on biological diversity, especially those species and ecosystems whose disappearance would likely do disproportionate harm to the ability of ecosystems to meet human needs or set in motion the extinction of many other species.

2. Hydrologic Forecasting

Recommendation: Establish the capacity for detailed, comprehensive hydrologic forecasting, including the ecological consequences of changing water regimes, in each of the primary U.S. climatological and hydrologic regions. Important specific research areas include all those described under Grand Challenge 4.

3. Infectious Disease and the Environment

Recommendation: Develop a comprehensive ecological and evolutionary understanding of infectious diseases affecting human, plant, and animal health.

4. Land-Use Dynamics

Recommendation: Develop a spatially explicit understanding of changes in land uses and land covers and their consequences.

IMPLEMENTATION ISSUES

The identification of grand challenges in environmental sciences and priorities for immediate research investment is only a prelude. The key then becomes implementation. In the committee's view, several critical implementation issues cut across all of the research areas identified. These issues include such matters as whether to proceed by establishing regional research centers, how best to support interdisciplinary research, and how to make environmental science useful to decision makers and managers and the public.

Recommendation: NSF, together with other agencies as appropriate, should conduct workshops that include research scientists in academia, the relevant agencies, and the private sector, as well as potential users of the research results, to discuss and plan research agendas and address implementation issues.

1

Introduction

Understanding our environment has long been central to the scientific enterprise, and is becoming increasingly important as growth in human populations and economic activities intensifies the stresses humans place on the environment. The consequences of those stresses are increasingly evident, such as habitat degradation; the hole in the ozone layer over high latitudes of the Southern Hemisphere; the increased rate of species extinction; changes in various elemental cycles in the soil, the air, and the oceans; and depletion of marine fish populations in many parts of the globe. This is also an exciting and challenging time for the environmental sciences. Progress in knowledge and theory has been stimulated by advances in computing power; in sensing technology below, on, and above the Earth's surface; in techniques and understanding of molecular biology that have increased our ability to understand ecological processes; and many other areas. In addition, there has been growing recognition of the value of multidisciplinary research involving natural and social sciences and engineering. Together these developments have led to a growing awareness of the central importance of the environmental sciences as humankind attempts to transition to a more sustainable relationship with the Earth and its natural resources. Advancing the environmental sciences, then, is both intellectually challenging and essential for the future of humankind. In this context, a key question arises: Of the many topics of great scientific excitement as well as great practical importance, which are the most important and urgent, and which are most likely to yield major results if tackled now? In other words, what are the grand challenges of the environmental sciences? The answer to that question is the topic of this report.

CONTEXT: THE MULTIDISCIPLINARY NATURE OF ENVIRONMENTAL SCIENCE

Most of the major challenges in the environmental sciences (and management) require multidisciplinary[1] solutions. The "environment" may be conceptualized in biological, chemical, physical, or social scientific terms, and important research endeavors arise from all these fields. New training, new organization, and new funding are needed to bring together multidisciplinary teams that can undertake research aimed at understanding the following:

- How natural systems[2] work.
- How human activities and other influences perturb these systems.
- What causes these perturbations.
- How changes in one system affect other systems and human well-being.
- How the knowledge needed to make well-informed choices about means of transforming or restoring environmental systems can be developed.

Natural systems—ecosystems; oceans; drainage basins, including agricultural systems; the atmosphere; and so on—are not divided along disciplinary lines; understanding any one of them requires expertise that cuts across several disciplines. For example, oceanic circulation patterns influence and are influenced by atmospheric circulation patterns, rainfall patterns, the topography of the ocean floor, temperature, and the chemistry of water, among other factors. Terrestrial ecosystems are affected by land use, land cover, and the climate system, as well as by the chemistry and biology of their constituent environments; while species within ecosystems are affected by physical-chemical inputs, population genetics, and interactions with other species, including humans. And because so many physical, chemical, and biological processes are strongly affected by and affect human activities, understanding those activities, including the development and use of technology, is integral to the environmental sciences. Thus environmental sciences include branches of social sciences and engineering just as they include branches of biological and physical sciences. For the environmental sciences to build the knowledge base they need, these disparate fields need to cooperate and collaborate.

Making science useful for environmental management is equally complex,

---

[1] By "multidisciplinary," the committee means a collaborative approach involving many disciplines; "interdisciplinary" implies integration of multidisciplinary knowledge. This usage conforms to the recent literature (e.g., Hansson 1999, Karlquist 1999, Policansky 1999).

[2] We use the term "natural systems" to refer to systems relatively undisturbed or not controlled by humans, as opposed to agroecosystems or urban areas. We recognize that humans and their activities are an integral part of many biophysical systems on the Earth and that to distinguish between human and natural systems is often a false dichotomy.

requiring a sound scientific and multidisciplinary understanding. Finding effective ways for scientists in a variety of disciplines to work together and to communicate with managers and governments is of great importance both for advancing scientific understanding and for making that understanding useful. It has often been difficult, however, to achieve the needed multidisciplinary collaboration, let alone interdisciplinary integration. The need to do so runs as a theme throughout this report and is implicit in the committee's recommendations. We return to this matter in Chapter 4.

STUDY PURPOSE AND SCOPE

The National Science Foundation (NSF) supports a wide variety of research in the environmental sciences, and as part of its long-range strategic planning sought advice from the National Research Council (NRC) about the most important and challenging scientific questions in the environmental sciences. NSF expects to use this guidance to help identify new research initiatives and programs that could move basic understanding forward in critical areas.

In response to NSF's request, the NRC established the Committee on Grand Challenges in Environmental Sciences. In recognition of the multidisciplinary nature of the subject under its charge, the committee was also asked to identify factors that may serve as barriers to the implementation of multidisciplinary research agendas, such as educational needs; research infrastructure, including equipment and institutional arrangements; and related matters. The committee consisted of 17 scientists drawn from a broad range of disciplines, including terrestrial and aquatic ecology, paleoecology, biogeochemistry, physical oceanography, biology, chemistry, physics, atmospheric sciences and climatology, hydrology, geology, environmental engineering, medicine, epidemiology, toxicology, geography, political science, economics, and psychology.

The committee's charge was not to identify grand environmental challenges (that is, to list the world's biggest environmental problems). Rather, it was asked to determine the most important research challenges within the environmental sciences, that is, areas of opportunity in which a concerted investment in science could yield new understanding. At the same time, these advances may also be relevant to understanding and solving the world's greatest environmental problems, given that scientific knowledge is a prerequisite for environmental problem solving.

APPROACH

Method of Soliciting Input

In identifying potential grand challenges, the committee made a concerted effort to obtain suggestions from as wide a sampling as possible of the scientific community and other interested individuals and organizations. The primary tool for soliciting these contributions was a letter inviting the recipients to submit a one-page description of a grand challenge for the committee's consideration (see Appendix A). This letter, which explained the purpose of the study and the criteria the committee would use to evaluate candidate grand challenges, was distributed to scientists throughout the United States and abroad via the e-mail listservs of dozens of professional scientific societies. Recipients were further invited to pass the invitation on to any others who might want to suggest potential grand challenges. The letter was also sent directly to scientists and managers at major federal government research agencies, to members of the National Academy of Sciences and the National Academy of Engineering, and to many NRC volunteers. An Internet site for the study was established and linked to the sites of scientific societies and other organizations so those who had not been contacted directly could learn about the study and submit their ideas to the committee.

This process for soliciting input generated more than 200 responses from people having a wide variety of backgrounds and affiliations with universities, governments, nongovernmental organizations, and the private sector in the United States and abroad (see Appendix B). Each submission was read and discussed by the committee. Many of the submitted ideas influenced the committee's deliberations and are reflected in the final list of grand challenges, although none of the submissions is included verbatim.

The committee also considered the results of earlier, similar exercises. These included many reports produced by the NRC, NSF, and others during the last decade that identified important research challenges within various disciplines and involving particular environmental issues.

Process for Selecting Grand Challenges and Immediate Research Investments

In response to the NSF request, the committee attempted to select a short list of high-priority research challenges. This strategy did not involve ranking environmental issues by importance, but evaluating opportunities for maximal research payoff. The committee developed its recommendations in two stages.

The committee first identified important broad areas of research, applying the criteria described below. This exercise resulted in eight grand challenges, along with the highest-priority substantive research areas for each.

The committee then selected four areas to recommend for immediate research investment. These selections resulted from a consideration of all eight grand challenges from the perspective of research implementation. The recommended areas are those the committee judges to have the highest likelihood of yielding a major payoff from increased investment in the next decade, given the current state of relevant science. These are not broad research recommendations addressed primarily to the scientific community, but actions that are intended to support scientific research and can be implemented by government officials, including NSF staff.

Finally, grand challenges in the environmental sciences may be different from other research activities in that they could require special efforts to develop measurement techniques, databases, or conceptual frameworks; to train scientists in new ways; to establish unusual collaborations among disciplines, universities, and government agencies; and the like. Accordingly, the committee considered these special needs with regard to the scientific enterprises selected as grand challenges.

The committee did not rank-order the grand challenges, nor did it rank-order the research recommendations, for the same reason: each of the challenges selected by the committee meets the above criteria, and each therefore deserves to be pursued vigorously by researchers and supported commensurately by research funders in the United States and worldwide during the next decade and beyond. Consequently, the order in which the challenges and research recommendations appear in subsequent chapters is simply alphabetical.

Selection Criteria

Grand Challenges

The committee agreed to select only a small number of grand challenges, even though there are many important and promising areas in the environmental sciences. By agreement with NSF, the committee considered what the most significant research challenges would be during the next 20-30 years. In other words, the committee focused on challenges that are likely to take at least one decade to engage successfully, in part to allow for the training of a critical mass of scientists to undertake the necessary projects. Although the committee did not exclude a priori challenges that could be met in a shorter time, the search favored longer-term scientific efforts.

The committee defined grand challenges substantively, that is, in terms of the kinds of knowledge to be developed. Although there are other sorts of challenges facing the environmental sciences—such as developing new methods and databases, training environmental scientists, and addressing mismatches between scientific needs and the structure of research organizations—we addressed such needs in the context of meeting substantive challenges rather than labeling any of them as grand challenges themselves.

The committee recognized that selecting a few grand challenges from an extensive list would inevitably be a somewhat subjective enterprise. To impose structure on its deliberations, the committee decided to use six criteria for selection:

• The challenge must be compelling. We selected only challenges we judged as offering the potential for a large payoff in both scientific and practical terms. Scientific payoff is of various types, including resolving important unanswered theoretical questions, opening new areas to systematic inquiry and explanation, and finding common explanations for phenomena previously believed to be unrelated. Practical payoff is also of various types, including the generation of useful information for avoiding or mitigating catastrophes, making long-term development plans, making economic choices in the face of environmental changes and uncertainty, and resolving public policy dilemmas.

• The challenge must be large, requiring numerous researchers, many years, and appropriate resources. Regardless of how important it might be, a challenge likely to be dealt with satisfactorily in a year or two of diligent, directed effort does not qualify as "grand."

• The challenge must be relevant to environmental issues of importance to humankind. Challenges were rated more highly if the research would address rapid environmental changes that are likely to require well-informed human responses in the near future, and if the environmental conditions under study would take a long time to correct if research revealed the importance of corrective action.

• A fourth criterion was feasibility. The committee favored topics on which research is likely to yield scientific payoff within a decade given the recommended level of effort, or on which an increased research effort now would help build the necessary knowledge base for important results later.

• The criterion of timeliness led the committee to emphasize topics on which research would be facilitated by recent developments in technology, data, theory, or scientific collaboration. Our reasoning was that breakthroughs are more likely in fields in which new tools or other capabilities have recently emerged than in those in which the existing research tools have already been in use for a considerable time.

• The committee favored challenges that require multidisciplinary collaboration. Challenges that might be met by research within a single discipline or research tradition were not ruled out. However, because multidisciplinary collaboration is both difficult and important for so much of the work in the environmental sciences, as discussed above, major research efforts that would build the capability for multidisciplinary collaboration would have positive spillover effects for the rest of environmental science, and therefore deserve priority.

Immediate Research Investments

To provide a shorter list of more focused recommendations for immediate research support, the committee reexamined the grand challenges and the focused research areas identified for each. We considered a dozen potential action items—research areas that met the above criteria for the grand challenges and could be recommended for immediate research investment. The potential action items outnumbered the grand challenges because the committee considered some research areas that cut across more than one grand challenge, and because each grand challenge encompassed more than one focused research area.

Choosing among the topics was itself a major challenge. Each topic had substantial merit based on the importance of the scientific questions involved and the potential benefit of increased study. Each could be characterized as timely, important, and even having some urgency. Therefore, the committee revisited its selection criteria and applied additional ones to narrow the list to three or four areas that would be recommended for immediate research investment. For each candidate topic, the committee asked whether the investment in that topic is especially timely, i.e., whether the time is ripe in terms of the balance between what is known and what is likely to be learned. The committee also considered for each topic the level of current research support in relation to the probable need for support. In other words, in identifying topics for immediate research investment, we ranked those we judged to be in need of significant additional funding higher than others, which were often deemed to be of equal intellectual and practical importance. The committee also favored those areas we judged to have potential for major and rapid progress. And we favored research areas for which we believed that significant research funding has the potential to transform disciplines by leading the development of new approaches and by encouraging cross-disciplinary interaction.

ORGANIZATION OF THIS REPORT

Chapter 2 describes the eight grand challenges identified by the committee, while Chapter 3 presents the committee's recommendations for immediate research investments. Chapter 4 addresses implementation issues.

2

The Grand Challenges

For each grand challenge described in this chapter, the committee judges that major scientific and/or practical payoff is likely to result if there is a significant infusion of research support over the next decade or two. We begin the discussion of each challenge by identifying the scientific payoffs that appear most likely and practical payoffs that the expected scientific advances would make possible. We then identify recent scientific progress that makes major advances in the area of the challenge possible now. Next we list focused research areas within each challenge that are especially deserving of intensive development. These lists are not intended to be comprehensive; rather, they include only those areas we judge most exciting and likely to yield major breakthroughs in the near future.

GRAND CHALLENGE 1: BIOGEOCHEMICAL CYCLES

The challenge is to understand how the Earth's major biogeochemical cycles are being perturbed by human activities; to be able to predict the impact of these perturbations on local, regional, and global scales; and to determine how these cycles may be restored to more natural states should such restoration be deemed desirable.

Practical Importance

Six nutrient elements—carbon, oxygen, hydrogen, nitrogen, sulfur, and phosphorus—make up 95 percent of the biospheric mass on the Earth and form the biochemical foundation for life (Schlesinger 1997). The cycling of these elements through the Earth system in their biological, geological, and chemical forms constitutes the biogeochemical cycles. Also included under the rubric of biogeochemical cycling can be elements such as potassium, calcium, molybdenum, iron, and zinc, which are needed as physiological regulators or cofactors for enzymes. Imbalance in the availability or utilization of these elements has both direct and indirect influences on the distribution and viability of many organisms.

Research during the last several decades has provided many insights into the importance of biogeochemical cycles. It is now recognized that the evolution of photosynthetic organisms more than 2 billion years ago transformed the Earth's atmosphere from strongly reducing to its current oxygen-rich state. The interrelationship between greenhouse gases and climate was identified more than a century ago (Arrhenius 1896). Today we understand that carbon dioxide (CO_2)-induced ocean warming was sufficient to trigger the large-scale destabilization of methane hydrates (Norris and Rohl 1999). This positive feedback with global effects occurred at the Paleocene/Eocene transition, and was associated with high-latitude warming and changes in terrestrial and marine biota. The concentrations of many greenhouse gases (e.g., CO_2, nitrous oxide [N_2O], and methane [CH_4]) have risen over the last 100 years at rates unprecedented in the geologic record. It is clear that these rapid rises in concentrations are being driven by global changes in the Earth's biogeochemical cycles. What is less clear is how long these changes in biogeochemical cycles will continue, what effects they are having on the climate system, how these effects will reverberate throughout the Earth system, and how positive and negative feedbacks within the system will interact to accelerate or ameliorate these effects.

Human actions strongly influence changes in the Earth's biogeochemical cycles, with potentially devastating effects. Combustion of fossil fuels and conversion of forested land to agriculture have redistributed carbon from plant, soil, and mineral pools into the atmosphere, where greatly increased CO_2 has the potential to alter climate, affect the photosynthetic efficiency of vegetation, and change large-scale ecosystem dynamics (Amthor 1995). The combustion of fossil fuels and the manufacture and use of nitrogen fertilizers have approximately doubled the annual supply of fixed nitrogen to the soil relative to preindustrial times, a circumstance that has the potential to alleviate nitrogen limitation of productivity in terrestrial ecosystems and may thus contribute to enhanced terrestrial carbon uptake (Holland et al. 1997). Similarly, ore smelting and coal combustion have roughly doubled annual emissions of sulfur gases to the atmosphere, with implications for both acid rain and global climate change (Galloway 1995). Anthropogenic perturbation of the cycle of phosphorus, a limiting nutrient for many plants, has been less studied, but is thought to be significant at least at a regional scale.

It is clear that these human-induced stresses to the biosphere interact, but the net effect of the multiple perturbations remains uncertain. Increased tropospheric CO_2 and widespread nitrogen deposition both act to fertilize plant growth, but other factors—such as soil acidification, high tropospheric ozone levels, loss of soil fertility through base cation loss, and their interactions with plant diseases and pests—all reduce plant productivity and have other effects on the biosphere. The net effect of these factors on crop productivity and the biosphere's ability to consume the carbon emitted through fossil fuel combustion needs to be understood. This is but a single example. We also know, for instance, that the current changes to the nitrogen cycle have had profound impacts on freshwater and perhaps oceanic resources and fisheries.

Human influences on the biogeochemical cycles are not all increasing so dramatically. Recent restrictions on sulfur dioxide (SO_2) emissions in some countries have resulted in reduced inputs of acid rain to surface waters and ecosystems. The production and emissions of chlorofluorocarbons (CFCs) have also been reduced. Despite these scientifically informed policies, however, the abundance of N_2O, CH_4, and sulfate aerosols, all biogeochemically important compounds, will interact with the changing climate to influence the rate of recovery of the ozone layer. Yet while the biogeochemical cycles of the nutrient elements constitute crucial constraints on the Earth's physiology, they remain poorly understood. This lack of understanding strongly limits our perspective on the many facets of global change. During the next century, continuing expansion of the influence of urbanization, industry, and agriculture on already perturbed biogeochemical cycles is likely. Increased scientific understanding of these cycles and the activities that are perturbing them is vital to formulating plausible political and social solutions to these important environmental perturbations.

Scientific Importance

The goal of biogeochemistry is to quantify the rates of transfer of relevant compounds and their accumulation or depletion in storage reservoirs. Knowing the residence time of compounds in each type of reservoir is central to predicting their changes over time. For example, during the last decade, research on the global carbon cycle has established that fossil fuel combustion has released an average of 5.5 (+/–0.5) gigatons (Gt) of carbon in CO_2 into the atmosphere each year, and land-use changes have contributed an additional 1.6 (+/–1.0) Gt, for a total of 7.1 (+/–1.1) Gt (Schimel et al. 1995). Only 3.3 Gt of carbon is actually stored in the atmosphere. Ocean uptake of 2.0 (+/–0.8) Gt leaves an additional 1.8 Gt to be accounted for—the so-called "missing sink" of carbon. The remaining carbon is probably stored on land, and the locations and mechanisms of this carbon storage continue to be the subject of discussion and research (Tans et al. 1990, Nabuurs et al. 1997, Fan et al. 1998, National Research Council 2000a, Schimel et al. 2000). The likely mechanisms are CO_2 or nitrogen fertilization of the biosphere; reforestation, resulting in carbon storage in wood; and interactions with climate and its interannual variability. Yet the lack of a complete understanding of the current carbon budget hampers efforts to understand past geologic changes and to predict future changes in CO_2 concentrations. The magnitude, global scale, and potential destructiveness of some cycle perturbations make research on these cycles particularly urgent and timely.

As indicated by its very name, biogeochemistry links scientific specialties. New discoveries have emerged as specialists in any number of areas have recognized that they must collaborate with scientists from other disciplines to solve their problems. Limnologists and oceanographers recognize that atmospheric chemists and ecosystem ecologists may be their best sources of information on future rates of nitrogen fixation. Researchers around the world are using the output of climate models to understand the internal dynamics of the ecosystems they study. Modelers, foresters, and botanists are beginning to appreciate how increases in nitrogen deposition may enhance carbon storage, for example, or how carbon uptake may be limited in other areas that are nitrogen-saturated (Townsend et al. 1996). Bringing these different perspectives together is important, but it poses a challenge for scientists and managers seeking to build workable structures that can support the needed science.

The ecosystem implications of the biogeochemical cycles come into focus most sharply when variations in space and time are taken into account. Ecosystems vary widely from place to place and over time for many reasons, and globally averaged cycle information relates only weakly to those unique situations. As the broad outlines of the biogeochemical cycles become better delineated, spatial distributions and temporal trends in the parameters of interest will link the cycles in increasingly useful ways to topics of interest within other grand challenges.

Scientific Readiness

The growth of the field of biogeochemistry during the past 10 to 15 years has led to significant theoretical and experimental developments that can serve as the base for future research, and the study of carbon and nitrogen cycles has greatly benefited from recent technological advances. Of particular note are analytical techniques for isotope analysis of ^{13}C, ^{18}O, ^{15}N, deuterium, and ^{14}C, as well as the measurement of an increasing array of atmospheric trace gases, including reactive oxides of nitrogen, sulfur gases, OH, and O_2. Direct flux measurements of energy, momentum, and CO_2 and H_2O vapor exchanges, not possible a decade ago, today have become a cornerstone of both the U.S. and European field experiment programs (Brasseur et al. 1996). Remote measurements of ocean and land surfaces and the atmosphere made possible by recent satellite launches (such as the National Aeronautics and Space Administration's [NASA]

TOMS instrument and Terra satellites) have and will continue to enable great advances in understanding. They will also fill gaps in the global information database, including the understanding of land-cover change argued for under Grand Challenge 7. Models have progressed dramatically, and are beginning to provide realistic simulations of the complex interactions among atmospheric, oceanic, and terrestrial systems (American Meteorological Society 1998).

The existence of long-term measurements made possible by funding from a number of federal agencies has been essential to progress in the field. These datasets include the global trace gas measurements made by the Climate Monitoring and Diagnostics Laboratory (1996-1997), funded by the National Oceanic and Atmospheric Administration, which have provided insights into the carbon cycle and carbon cycle models. NASA's archiving of Landsat satellite images has enabled quantification of large-scale land-use change (Skole and Tucker 1993). The Environmental Protection Agency's surface observations of pollutants and the development of emission inventories have helped test our understanding of atmospheric chemistry (Guenther et al. 1994, Benkovitz et al. 1996). The National Atmospheric Deposition Program/National Trends Network Program and the National Dry Deposition Network have provided long-term measurements (1978-present and 1990-present, respectively) of wet and dry deposition that enable regional and national evaluations of acid rain inputs, nitrogen deposition (Holland et al. 1997), base cations inputs (Driscoll et al. 1998), and surface water resources. The Department of Energy's funding of the Carbon Dioxide Information and Analysis Center has provided a much-needed synthesis of CO_2 data at a critical time. Maintaining these long-term data programs is seldom easy, but is crucial to deriving increased insight. The above are but a few key examples of successes in the field.

We are now poised to place our understanding of biogeochemical cycles on a much firmer theoretical and empirical base than now exists. In the coming decade, it will be possible to gain a solid quantitative understanding of the cycles and budgets of the key biogeochemical constituents. In fact, a well-developed strategy (the U.S. Carbon Cycle Science Plan) already exists for understanding the cycling of CO_2. Continuing major commitments of financial and human resources by multiple agencies are needed to bring this plan to fruition. An ultimate goal is to make reliable predictions of future changes in these cycles and the resulting effects on planetary functioning. Progress toward this goal will depend on continued research on biogeochemical processes and on human activities that drive these processes. (The extent to which this approach spans disciplinary areas is indicated by the fact that the use of the nutrient elements and of land, water, and various natural materials is addressed in Grand Challenges 7, 4, and 8, respectively.) In a policy context, predictive biogeochemical models could help guide decisions about such matters as fossil fuel use, energy production, agricultural and industrial practices, and mitigation of climate change.

Important Areas for Research

1. Improve the quantification of sources and sinks of the nutrient elements, and gain a better understanding of the biological, chemical, and physical factors regulating transformations of nutrient reservoirs. Greatly improved estimates of the sizes of nutrient reservoirs on regional and global scales and their rates and causes of transformation are essential for identifying those reservoirs and transformations most influenced by human activity and predicting the impact of the transformations on ecosystem health; global climate; and human needs, such as food supplies and clean air. Studies of the Earth's history can reveal the significance of biogeochemical cycles in altering climate and the distribution, abundance, and diversity of organisms, and aid in understanding positive and negative feedbacks within the global system.

2. Improve understanding of the interactions among the various biogeochemical cycles. Nitrogen, phosphorus, and essential trace nutrients such as iron alter the productivity of terrestrial and oceanic plants and the transfer of carbon from the atmosphere to living organisms. Likewise, decomposition and remineralization of organic matter transform nutrients captured by organisms back into inorganic form. All of the cycles of essential nutrients interact with each other, and the positive and negative feedbacks among them are at present poorly quantified and understood. In addition, the biogeochemical cycles are strongly influenced by the terrestrial hydrologic cycle. An understanding of these synergisms and their impacts is necessary if changes in any one cycle are to be predicted.

3. Assess the impacts of anthropogenic perturbations of biogeochemical cycles on ecosystem functioning and atmospheric and oceanic chemistry, and develop a scientific basis for societal decisions about managing these cycles. Greatly improved projections of future concentrations of CO_2, CH_4, nitrous oxides, and aqueous and atmospheric pollutants, as well as understanding of the responses of natural and managed ecosystems to these and other atmospheric components, are required to make wise management decisions regarding human activities. Better projections will depend on research to improve understanding of the drivers of human actions that perturb the cycles and to enhance models of biogeochemical processes and their ecological effects. An understanding of the impacts of past and current land-use and agricultural, industrial, and domestic practices and policies on nutrient cycles would facilitate the development of models for fully assessing those impacts. In addition, the cycles of non-nutrient elements, addressed in Grand Challenge 8, Reinventing the Use of Materials, are important to ecosystem functioning. Thus a longer-term goal is to integrate the environmental implications of the nutrient and non-nutrient elements. Research on the effects of changes in biogeochemical cycles on human societies and economic activities is also an essential part of the scientific basis for societal decisions.

4. Explore prospects for mitigating these perturbations. There is a need for extensive research regarding the feasibility and effectiveness of a variety of both technical approaches (e.g., precision agriculture, creation of carbon sinks, technologies for more efficient uses of nutrient elements) and institutional approaches (e.g., financial incentives for resource conservation, creation of emissions markets) for achieving sustainability of the essential nutrient cycles. This research priority has obvious overlap with Grand Challenge 6 on institutions and resource use.

The research priorities for biogeochemistry are clearly related to those for a number of the other grand challenges in addition to the overlaps noted above. Significant changes in biogeochemical cycles are often driven by extreme weather events, such as those outlined in Grand Challenge 3 on climate variability. Moreover, it is clear that interannual variation in climate drives interannual changes in carbon and possibly nitrogen cycling (Braswell et al. 1997, Erickson 1999). Understanding the linkages between micronutrient and nutrient cycles, as well as transforming that understanding into meaningful policy, will also require information and insights gleaned from Grand Challenge 3. Vitousek et al. (1997b) have shown how acceleration of the nitrogen cycle can affect biodiversity and species composition in terrestrial and aquatic ecosystems, effects that have obvious overlap with Grand Challenge 2 on biological diversity and ecosystem functioning. In addition, acceleration of the nitrogen cycle is implicated in the widespread hypoxia in the Gulf of Mexico, in freshwater pollution following the North Carolina floods of 1998 and 1999, and in the *Pfiesteria* outbreaks along the Eastern Coast of the United States, addressed by Grand Challenges 5 and 6 on infectious disease and institutions, respectively. And changes in land-use dynamics (Grand Challenge 7) have driven large-scale changes in the carbon and nitrogen cycles.

GRAND CHALLENGE 2: BIOLOGICAL DIVERSITY AND ECOSYSTEM FUNCTIONING

The challenge is to understand the regulation and functional consequences of biological diversity, and to develop approaches for sustaining this diversity and the ecosystem functioning that depends on it.

Practical Importance

Human impacts on the land and oceans are pervasive and profound. The human enterprise has appropriated nearly half of the Earth's primary productivity, more than doubling the global cycling of nitrogen (Vitousek et al. 1997a,b). Humans harvest much of the oceans' production as well, drill petroleum from continental shelves, and are poised to begin using the deeper ocean floors for both mining and waste disposal and petroleum recovery.

Human use of an area has generally meant its severe degradation as a natural habitat. Ecosystems and their functioning are threatened. As a result, the rate of species extinction is higher now than at almost any time in the Earth's history (National Research Council 1995). Today, indeed, we face the risk of a great mass extinction, one of only a handful in the history of the Earth.

The permanence of extinction makes it qualitatively different from other kinds of environmental change. Many societies around the world support the protection of species diversity, often explicitly, on ethical, moral, cultural, and aesthetic grounds. Many U.S. federal and state laws support the maintenance of species diversity. For example, the U.S. Endangered Species Act of 1973 states that it is "the policy of Congress that all Federal departments and agencies shall seek to conserve endangered species and threatened species. . . ." (Section 2 {b[c]}). Thus an anthropogenically driven mass extinction would be a great societal as well as biological loss.

Such a loss would also be risky. Humans depend crucially on nature for many things, from food, fiber, and medicines to recycling of nutrients and regulation of air quality, water quality, and climate (Daily 1997, National Research Council 1999f). Environmental scientists do not yet fully understand the sensitivity of these things to changes in the diversity of organisms and ecosystems. At present, we have a limited appreciation of what is really at risk, of the time scale for losses, and of the environmental consequences of simplifying and mixing the Earth's biota. Nonetheless, a major loss of biological diversity clearly threatens the capacity of the Earth to support human societies.

To predict the impacts of human activities on the diversity of genotypes, species, and ecosystems, we need a thorough understanding of the fundamental natural controls on biological diversity. We also need to make a major investment in discovering to what extent ecosystems with altered diversity can provide the services humanity depends on. Further, progress made in understanding the genesis and regulation of biological diversity needs to be applied in developing the capacity for preserving that diversity. Given the already pervasive impacts of human activity, high priority must be placed on the formulation of strategies for integrating conservation with human uses.

Threats to biological diversity on the land and in the oceans are generally unintended consequences of the development of human societies, growth in human populations, and efforts to improve standards of living. Practical efforts to protect species and ecosystems must reconcile ecological objectives with human needs.

Scientific Importance

Throughout its history, the field of ecology has focused on understanding the factors that produce and control biological diversity (e.g., von Humboldt 1807, Preston 1948, Hutchinson 1959, Rosenzweig 1999). Success would be a substantial intellectual prize. It would represent a pinnacle of knowledge of the Earth's living systems—comparable to the goal of cosmology to discover the events and processes that determine and guide the development of the physical universe. The practical value of such understanding would appear to be inestimable.

Since the early 19th century, observers have noted striking variations in patterns of species diversity with latitude, productivity, climate, and area (e.g., von Humboldt 1807). Area and isolation are fundamentally important, reflecting control of local and regional diversity on shorter time scales by the balance between migration and local extinction. Understanding of the relationship between species diversity and area—known as a species-area curve—is a powerful tool. At longer time scales, speciation also becomes important as the factor generating species diversity.

Although considerable understanding of the processes that lead to new species and those that destroy established ones has been achieved (e.g., National Research Council 1995), we do not yet know how to fuse that understanding into a quantitative theory capable of predicting changes in continent-scale or even local species-area patterns. It is not yet known whether a local extinction in one group will cause extinctions in others or whether species introductions, which are such an important part of the modern biological landscape, always lead to compensating or amplified losses in the diversity of native species. Without quantitative theories, we have only limited ability to predict rates of change or specific losses and gains that will follow a perturbation in the environment. However, current theories can be applied successfully to rank species diversities both within and among scales (MacArthur and Wilson 1967, Rosenzweig and Ziv 1999). Thus, a concerted effort during the coming decade could bring substantial advances.

Meanwhile, recent deep-sea research has taught us that the planet's deep ocean floor—most of the Earth's surface—harbors many more species than was previously believed. Thus, many of the species of the deep sea and their patterns of diversity remain to be discovered. At present, we do not know even the major features of the biogeography of the deep sea. The technology needed to obtain this information now exists. But the vastness and severe habitat of both the abyss and the edges of the continental shelves make sampling expensive and have restricted such activities. At present, deep-sea habitats remain wildernesses, and as such they allow the study of diversity in an environment relatively unaffected by human activities. Soon they may be affected by petroleum drilling, mining, waste disposal, and fishing. An infusion of major support is therefore needed to take advantage of the current window of opportunity.

Diversity in terrestrial soils is poorly characterized as well. Although soils are easier to study in many ways than the deep sea, what their diversity means for microbes is not well understood. Because microbes are such an old and large fraction of the Earth's biota, improving this understanding is of great scientific and practical interest.

It is also important to understand diversity at scales larger and smaller than that of the species. Past changes in the number and distribution of the major terrestrial biotic communities, or biomes, are important keys to understanding the history of the Earth. Understanding the limits on the number and distribution of biomes becomes more critical as human-caused climate change creates pressures for biome shifts and perhaps for the disappearance of some biomes and the emergence of others. Dynamic global vegetation models (International Geosphere-Biosphere Programme 1997) are a recent attempt to simulate the number, diversity, and distribution of biomes, based on competition among plants representing the major functional types. Unfortunately, our understanding of the factors that control this competition is still limited, and the results of these models are therefore tentative.

The study of the relationship between biological diversity and ecosystem structure and functioning is in its infancy. Early studies have produced many examples but few general principles (Tilman 1999, Wardle et al. 2000, Naeem 2000). Obviously, at the lower limit (only one or very few species), loss of species diversity must affect ecosystem functioning, but there is no general principle concerning the impact of decreasing biological diversity on the risk of widespread loss of ecosystem functioning. It is clear that not all species are equally important, but little is known about the general extent to which ecologically similar species can substitute for each other in providing ecosystem services. A dedicated effort combining experiments with long-term studies, opportunistic observations, and synthesis would greatly advance understanding of the relationships between diversity and functioning. Although we cannot predict the results of these studies, almost any result would be of great value. Whether there is a general relationship, no relationship, or—most likely—different relationships under various circumstances, the knowledge will be essential for understanding and preserving biological diversity and ecosystem services.

For much of the 20th century, researchers in population genetics and population biology sought to understand the factors that regulate a third scale of biological diversity—the genetic diversity within species and populations. Biologists succeeded in many particular cases. But they lack a comprehensive theory linking genetic diversity with other factors, including environmental stresses and diversity at the level of species or ecosystems. While it is clear that genetic diversity is a powerful influence on ecological success and hence on the persistence of species, we cannot yet quantify this relationship, although many examples illustrate the vulnerability of low-diversity agricultural systems to attack by pests.

Even total understanding of the laws of diversity would be inadequate by itself to conserve diversity in the face of the changes humans make in the environment. Research is also necessary on the needs of specific species and ecosystems that have been truncated by human activities. Moreover, as noted above, practical efforts to protect species and ecosystems must be based on a balance between conservation needs and human needs. Achieving such a balance will entail answering many scientific questions related to three major strategies for protecting biological diversity—reservation, restoration, and reconciliation:

- *Reservation* is the setting aside of natural and near-natural areas for non-human biota. This strategy, exemplified by the establishment of national parks, has grown into a U.S. and worldwide program. It has reduced species losses, but it has not and cannot by itself eliminate them because so much natural habitat has been altered by human activities. Nonetheless, research is important to improve the design and implementation of biological reserves.
- *Restoration* ecology—only now beginning to see large-scale scientific application—attempts to return degraded sites to some degree of natural structure and functioning (see, e.g., National Research Council 1992). Restoration has much to offer for protecting biological diversity but is challenging, largely because of incomplete knowledge of which aspects of an ecosystem must be restored to protect an endangered species and to what degree of functioning. For example, Zedler (1996) describes how an apparently successful restoration of the vegetation in a coastal wetland did not support endangered clapper rails because the cordgrass was not tall enough to support their nesting. Similarly, red-cockaded woodpeckers do not depend simply on the presence of long-leaf pines, but require nest-holes in living trees (McWhite et al. 1993). And natter-jack toads need more than early successional stages of sandy heathlands; they must have ponds warm enough to support early breeding so their tadpoles can escape predation by tadpoles of the common toad (Denton et al. 1997).
- *Reconciliation* ecology is beginning to emerge as a scientific discipline. Reconciliation is based on the premise that there are ways to design and manage habitats for productive human use and the maintenance of natural biota.

Given continued human dominance of most terrestrial ecosystems, successful conservation of biological resources will depend on continued advances in our understanding of reservation, restoration, and especially the relatively new field of reconciliation ecology.

Scientific Readiness

The following conditions make a scientific initiative on biological diversity and ecosystem functioning particularly timely.

Advances in understanding biogeography, speciation, and extinction. Species-area patterns are now known to exist at four scales, each of which has been associated with a set of processes ranging from sampling artifacts, to co-evolution, to speciation-extinction dynamics. Many details of these relation-ships and of their mechanisms of action are beginning to emerge, creating the opportunity for a concerted effort to combine the understanding of these processes into a quantitative theory of species diversity.

Progress in understanding the interaction of biodiversity and ecosystem functioning. Many recent studies have explored aspects of the relationship between the diversity of species (and in some cases the diversity of genotypes or ecosystems) and ecosystem functioning (e.g., Chapin et al. 1997, Tilman 1999). The focus of these studies ranges from primary production, to resistance to biological invasion, to leakiness for nutrients. Given the support of extensive experimental and observational work, the next decade or two could see the emergence of a general theory. Even if it were discovered that there is no general relationship, that information would be of enormous scientific and practical value, so this work is certain to produce important results.

The idea that diversity itself causes evolution (Cody 1975) has led to the investigation of several crucial issues, including the evolution of specialists (Brown and Pavlovic 1992) and the factors that influence how rapidly and strongly evolution occurs in response to diversity (Holt and Gomulkewiecz 1997). Incorporation of such coevolutionary theories into experimental work on ecosystems and diversity and into ecological models will improve understanding and the accuracy of predictions.

New and improved tools. Several tools with direct relevance to the study of diversity have substantially improved the pace and quality of diversity research:

• *Satellite remote sensing* yields global maps of ecosystem distribution at a spatial scale of 1 km, and even higher resolution will be available soon. Satellite sensing reveals the distribution and diversity of ecosystem types, information critical for testing and improving models of ecosystem diversity.

• *Deep-sea sampling* routinely produces cores from both medium and abyssal depths using remotely controlled submersibles, providing new methods of assessing and understanding patterns of species diversity.

• *Genomics* using polymerase chain reaction (PCR) and microarrays can now be used for rapidly and efficiently assessing genotypic diversity and variation in gene expression. Molecular tools for characterizing microbial diversity reveal vast stores of hidden diversity in oceans, sediments, and soils, including environments at extremes of temperature and pressure. These methods will lead to new insights into the significance and consequences of diversity below the species level, as well as better understanding of species diversity.

• *Dynamic global vegetation models* integrate the results of research on plant ecology, soil, water, and atmospheric conditions. They attempt to forecast changes in plant cover in response to environmental variations.

• *Bias-reduction software*, based on work by Burnham and Overton (1979) and Chao and Lee (1992), reduces by an order of magnitude the sample-size biases associated with diversity estimation (Chazdon et al. 1998, Turner et al.

2000). Thus it dramatically advances our ability to perform fast, reliable assessments of species diversity and measure the dynamic responses of diversity to environmental changes.

Progress in conservation science. There is a large and growing store of experience with restoration of relatively small habitat patches and with the design of terrestrial and marine reserves (National Research Council 2001). Meanwhile, conservation science has increasingly turned its attention to the land humans continue to use. Various investigations have shown that human use of an area does not have to preclude its use by other species. Habitats for human use—if exploited with care—can harbor large numbers of native species (Daily 1999). Conservation ecologists are learning how to modify land-use techniques to favor diversity. They are also discovering that—for surprisingly small investments—they can adapt human landscapes to sustain target species that may be imperiled (e.g., Yosef and Grubb 1994).

Integration of ecology with economics, psychology, and sociology. Cultural, economic, and psychological factors drive human actions, mediate human preferences for environmental conditions, and thus help shape the configuration of landscapes in which diversity must survive. The relationship between the disciplines that study these factors and ecology is not well developed, but it is an increasing area of focus, with new programs, journals, and paradigms emerging. Many challenges remain before these disciplines are effectively integrated, but the right conversations are now under way, and future progress should be rapid.

Important Areas for Research

1. Improve tools for rapid assessment of diversity at all scales—species, population, and ecosystem. New technologies—such as molecular techniques and remote sensing—should be incorporated in such tools as they are required and become available. Continuing work is also needed on the development of techniques for assessing diversity from incomplete sampling and on the use of remotely sensed data to examine ecosystem characteristics.

2. Produce a quantitative, process-based theory of biological diversity at the largest possible variety of spatial and temporal scales. The goal should be to predict the diversity of biomes, growth forms, and functional types, as well as species and genotypes. To attain that goal, it will be necessary to continue to investigate and interrelate species-abundance and range-size distributions, population structures and densities, and productivity patterns, as well as mechanisms of speciation and the relationship of population size to evolutionary change. In addition, theories of coevolution need to be extended and matured.

3. Elucidate the relationship between diversity and ecosystem functioning. Much evidence suggests that biological diversity affects ecosystem functioning mainly through sampling: a high level of diversity increases the probability of including a species or functional type that fills a particular role. It is not known whether this is true in general, or other principles become more important under some conditions. In short, there is no theory for the role of diversity in ecosystems. A series of experiments is needed to test explicit hypotheses about the mechanistic controls on biological diversity at all scales, and about the relationship between biological diversity and ecosystem functioning, including persistence. Some of these experiments would involve manipulated diversity and landscape complexity. Others would involve the consequences for diversity of a range of patterns of human activity.

4. Develop and test techniques for modifying, creating, and managing habitats that can sustain biological diversity, as well as people and their activities. Such work would depend on having an understanding of the design of the habitats in which people live and work, as well as the factors that influence human choices and preferences for different habitat types. Much of the science required would reveal the habitat requirements of imperiled species and the degree to which their continued existence is required for the adequate functioning of an ecosystem. It would also involve identifying species whose loss would likely lead to a cascade of further extinctions. As part of this effort, it would be necessary to develop management techniques that could be used to keep spatially diminished ecosystems at work.

GRAND CHALLENGE 3: CLIMATE VARIABILITY

The challenge is to increase our ability to predict climate variability, from extreme events to decadal time scales; to understand how this variability may change in the future; and to assess its impact on natural and human systems.

Practical Importance

Although climatic changes have occurred throughout the Earth's history, the accumulation of greenhouse gases in the atmosphere is perturbing the climate system with unknown effects on climate extremes and variability. The increase in greenhouse gases may be affecting the frequency and magnitude of severe events (e.g., episodes of heavy rainfall) (Intergovernmental Panel on Climate Change 1996) and may also be changing seasonal weather patterns (e.g., length of growing season, number of snow days, duration of ice cover on lakes). Because human land use has altered the resiliency of natural ecosystems, changes in weather extremes and in interannual variability may have a larger impact on ecosystems than the increases in average temperatures projected for the next century. Indeed, even if the spectrum of extreme events and climate variability were to remain unchanged, the impact of droughts, floods, and severe storms would probably increase as a result of extensive human alteration of landscapes through removal of forests, drainage of wetlands, channelization of rivers, construction of cities on floodplains, and growth of human populations in high-risk coastal areas. Mortality and morbidity associated with temperature extremes, loss of livestock confined in feedlots or barns under extreme conditions, extensive property damage due to hurricanes such as Andrew and Floyd, and significant agriculture losses and water supply problems during drought have substantially increased public awareness of the safety issues and economic impacts of extreme weather events. Yet despite the importance of these issues, understanding of how climate variability is likely to change in response to global warming and large-scale land-use changes remains poor.

Scientific Importance

Several factors underlie the scientific importance of this research challenge. First, although we are far from a comprehensive theory of climate variability, substantial progress has been made in understanding some of its aspects, in particular El Niño-Southern Oscillation (ENSO) events, a major source of climate variability on seasonal to interannual time scales. Observations recorded in ice, corals, and tree rings are extending the historical record of ENSO, and new observational and modeling capabilities have greatly improved our ability to predict the evolution of ENSO events. However, comprehensive predictions that capture the extent of ENSO and its impacts in different regions of the Earth are not yet possible. As a consequence, our ability to assess changes in ENSO over the next century is highly limited. Research on other modes of climate variability is in its infancy, and characterizing those modes is necessary if we are to unambiguously discern long-term climate trends caused by human activities, as well as understand natural variations in the global carbon cycle.

Second, many investigators believe the intensity and frequency of extreme events, including hurricanes, ice and snow events, floods, and droughts, change significantly in concert with longer-term climate changes. Investigators have found recent changes in the character, frequency, and seasonal patterns of extreme weather events (Intergovernmental Panel on Climate Change 1996), raising concern that these patterns are due to anthropogenically driven climate change. Yet the mechanisms controlling these variations are still largely unknown, and extreme events remain among the most uncertain of all climate projections.

Third, fine-resolution sampling of paleorecords reveals sudden shifts of climate occurring within years or decades at many different times in the past, suggesting that the climate system can shift from one mode to another as certain thresholds are crossed. There is concern that the rapid changes in climate in the coming century could trigger such a shift. Yet our understanding of the mechanisms of abrupt climate transitions is extraordinarily limited.

Fourth, extreme events and climate variability have dramatic potential to alter coastal and terrestrial ecosystems through direct temperature, precipitation, and physical (e.g., wind) impacts; through changes in freshwater inputs; and through indirect changes to air and water quality. The potential impacts extend from habitat disruption to changes in species composition and diversity. Yet we understand very little of how changes in the frequency and character of events or in interannual variability may influence ecosystems.

Finally, human societies and economic systems have adapted to historic patterns of climate variability, but may be disrupted to various degrees, depending on their coping capacities, if these patterns change. In addition, human alterations of the landscape may have changed the vulnerability of social systems to climate variations within historic ranges. Understanding the potential impact on humans of changes in patterns of climate variation depends on improved fundamental understanding of such human-climate interactions.

Scientific Readiness

Comprehensive models incorporating the atmosphere, oceans, vegetation cover, ice, and biogeochemistry are required to assess the nature of the climate changes associated with various factors, most of which are themselves rapidly undergoing change due to human activities. Newer, faster computers are driving the development of comprehensive interactive models with coupled atmosphere and oceans. However, because of limited access to powerful supercomputers, U.S. modeling centers have found it difficult to perform high-resolution studies of coupled ocean-atmosphere climate change, and this in turn has hampered scientific progress in understanding fundamental climate processes (National Research Council 1999b). Vegetation has large effects on climate. Although incorporating these effects into climate models introduces significant challenges, important work has begun. Since human land use may alter vegetation more rapidly than natural processes, collaborations between social and natural scientists become essential so that reasonable predictions of human behavior can be incorporated into the models. Such collaborative modeling studies have shown great potential, although examples are still few.

The finer spatial scale of recent models increases their potential for simulating weather events and patterns of variability on short time scales that can be verified through comparison with the rapidly expanding observational record. Remote sensing of vegetation and a variety of key climatic variables, combined with weather and deep-ocean observations, provide a much greater capability to study climate and climate change. Paleorecords from many parts of the globe are beginning to provide a strong foundation for comprehensive models of the Earth system. Fine-resolution records from ocean sediments, ice cores, and lakes are making it possible to describe levels of climate variability on annual and decadal scales and to recognize extreme events. All of these new records enable testing of climate models to assess their ability to predict climate variations under changed boundary conditions. Finally, multidisciplinary investigations are beginning to allow ecologists to interact with social scientists in examining both ecological and human responses to climate change and variability.

Important Areas for Research

1. Improve observational capability. As noted by the National Research Council (1999a), our instrumental capacity to observe the Earth's climate system is deteriorating worldwide, greatly limiting the ability to adequately document climate variations. It is crucial to strengthen and revitalize these observational systems, originally designed to monitor weather, so we can better understand the spatial and temporal attributes of climate variations. Long-term, consistent, and accurate observations are needed, along with enhanced observations of climate-related ecosystem and social phenomena and the ability to take advantage of technological advances, such as new satellites and ocean monitoring systems.

2. Extend the record of observations. Historical observations of climate represent only a small segment of time and are inadequate for assessing the nature of climate variability. Paleorecords are being dated more precisely, and high-resolution data are being compiled for a variety of indices. Enhancing the quality of these observations and extending the records spatially and temporally are critical to a full understanding of climate variability. Where possible, paleo-records should overlap with the instrumental record, enabling the development of integrated historical and proxy datasets. Paleorecords of climate should be linked with paleoecological, archaeological, and historical data to build a basis for improved understanding of climate interactions with ecosystems and social systems. Doing so would also improve our capability for hydrologic forecasting (Grand Challenge 4).

3. Conduct diagnostic process studies. Uncertainties in our understanding of climate variations and interactions among climate, ocean circulation, carbon cycling, atmospheric chemistry, vegetation, hydrology, and human systems should guide focused field and model studies. These studies should be directed at the controls of climate variability; they must include an emphasis on boundary layer processes, linkages among the ocean-atmosphere-land surface, more explicit representation of climate-vegetation interactions, evaluation of ecosystem implications, and process studies of human coping mechanisms. Portions of this research overlap strongly with Grand Challenge 1 (Biogeochemical Cycles), Grand Challenge 2 (Biological Diversity and Ecosystem Functioning), and Grand Challenge 7 (Land-Use Dynamics).

4. Develop increasingly comprehensive models. Neither a predictive understanding of climate variability nor an assessment of the interactions between climate and other critical elements of the Earth system is possible without the development of increasingly comprehensive coupled models. Greater attention should be given to (a) model-data and model-model comparisons, with an emphasis on testing these models against known geologic evidence and observed climate variations; (b) elimination of major uncertainties in model parameterizations; and (c) the development of predictions at spatial and temporal scales as appropriate for the examination of biologic, hydrologic, and socioeconomic systems. Portions of this research overlap with Grand Challenge 4 (Hydrologic Forecasting).

5. *Conduct integrated impact assessments, and study human responses to climate change.* A key challenge is to understand how climate variability interacts with terrestrial, freshwater, and marine ecosystems; water and food supplies; and the quality of human life. Improved prediction of climate variability is insufficient without careful assessment of the impacts of climate variability and a much greater understanding of the linkages between climate variability and natural ecosystems. Also needed is improved knowledge of human responses to a changing climate (e.g., changes in land use), which themselves can have major environmental effects.

GRAND CHALLENGE 4: HYDROLOGIC FORECASTING

The challenge is to predict changes in freshwater resources and the environment caused by floods, droughts, sedimentation, and contamination in a context of growing demand on water resources.

Practical Importance

Water is an essential natural resource that shapes regional landscapes and is vital for ecosystem functioning and human well-being. Human use and contamination of freshwater are stressing the resource, and alterations in the hydrologic regime have serious consequences for people and the environment. This grand challenge addresses the need to forecast both the hydrologic regime and the environmental consequences of changing that regime.

Human use of fresh water. In the next two decades, water use is expected to triple in the world (L'vovich and White 1990, Postel 1998), leading to corresponding increases in pollution, erosion, runoff, dewatering, and salinization. Although per capita domestic water use in the United States is 500-600 liters per day, total daily per capita water use in urban areas is about 5,000 liters (Solley et al. 1998). To satisfy the growing demand for water, the United States has built more than 75,000 dams (Graf 1999) and has exploited groundwater resources to the extent that major aquifers are being mined and the resource consumed (Graf 1993, Bredehoeft 1984). During the last few decades, depletion of aquifers has also become a widespread problem in parts of China, India, North Africa, and the Arabian peninsula, leading to critical water shortages, especially among poor, rural communities (Postel 1999).

Threats to freshwater ecosystems. Human demands on water resources have strong effects on the integrity of freshwater ecosystems (Naiman et al. 1995, Naiman and Turner 2000). In the United States, only about 2 percent of the 5 million km of streams is in good condition, and more than half of the animal species listed federally as threatened or endangered are aquatic. Nationally, 39 percent of native fish species are rare to extinct, and many others have a high to moderate risk of extinction in the near future (Stein and Flack 1997). This situation is due mainly to hydrologic alterations of freshwater habitats and to the presence of introduced, nonnative species. Similar ecological stresses are occurring in many other parts of the world, where major river systems, such as the Nile in northeast Africa and the Ganges and Indus in southeast Asia, have been heavily altered by dams, reservoirs, and diversions (Postel 1999).

Social and environmental impacts of floods and droughts. From 1990 through 1997, floods were responsible for more than $34 billion in damage in the United States alone (National Drought Mitigation Center 1999). In poor countries whose populations are highly vulnerable to weather disasters, the impacts of floods can be enormous. When record flooding occurred in the Yangtze River basin in China in 1998, more than 2,000 people drowned, and millions were driven from their homes. Prodigious floods occurring in Southern Africa in February 2000 displaced several hundred thousand people in Mozambique, Botswana, South Africa, and Zimbabwe. Damage due to drought is more difficult to quantify, but agricultural losses (and in poor countries, resulting problems of malnutrition) can be severe. The magnitude of the impacts of floods and droughts is a function of both hydrologic processes and human interaction with the environment.

Consequences of water contamination. Point- and non-point-source surface water and groundwater contamination threatens human health and natural ecosystems. Cleanup cost is one measure of the magnitude of the problem. The Environmental Protection Agency (1998) has estimated that there are 217,000 point-source sites in the United States, most of which affect groundwater, and that it will cost about $187 billion (in 1996 dollars) to clean them up. The use of pesticides and herbicides has led to widespread soil and groundwater contamination. For example, of 45,000 wells around the United States tested for pesticides, 5,500 had harmful levels of at least one.

Scientific Importance

Currently, our understanding and predictive ability with regard to hydrologic forecasting are limited by theory, method, and the scope of available models, as well as by data. Recent and evolving developments in remote sensing of parameters such as precipitation, soil moisture, snowpack, river discharge, vegetation cover, and surface topography are beginning to yield spatial and temporal data that are driving a revolution in hydrologic science, making it possible to measure hydrologic phenomena never before seen and thus poorly understood.

Yet the hydrologic and ecological theory, methods, and facilities needed to take advantage of this high-resolution information do not exist. Theoretical and methodological advances in hydrologic science are therefore needed to use and interpret field measurements and the abundant remote sensing data that soon will become available. A sustained research effort is likely to result in major advances in interpreting the behavior of hydrologic systems across different spatial and temporal scales; forecasting changes in water quantity and quality; and determining the impacts of these changes on surface and subsurface water resources, landscape dynamics, ecological communities, and human systems. These points are elaborated below in the discussion of important areas for research.

In meeting this challenge, science would draw on new high-resolution atmospheric, surface, and subsurface data obtained as a result of rapid advances in remote sensing and geophysical technology. Multidisciplinary collaboration, field measurements and experiments, and data integration would enable the development of a new body of hydrologic science, linking traditional hydrology, geomorphology, and aquatic/riparian ecology.

Scientific Readiness

The primary obstacles to advances in hydrologic research have been limited, sparse, spatially distributed data and broad disconnects between the scales of data generated. Recent and projected technological advances in remote data collection, coupled with field experiments, can supply abundant information about vast regions of the Earth at increasingly finer spatial and temporal scales. These data—including high-resolution visual, radar, and infrared satellite-based maps of the land, water, and atmosphere; precise surface topographic maps; new geophysical images of the shallow subsurface; and real-time, integrative environmental information—have never before been available. (For details on specific sensors and monitoring techniques, see National Aeronautics and Space Administration 1999a,b; National Research Council 2000b.) When linked with data on human consumptive use of water, contaminant emissions, and land-use patterns, this new information will provide the basis for greatly improved understanding and prediction of hydrologic and related environmental processes.

For the past three decades, hydrologists have built quantitative process-imitating models of water flow, sediment transport, channel dynamics, and contaminant migration. However, their ability to make hydrologic and ecological forecasts has been limited by the lack of understanding of interactions across multiple spatial and temporal scales. The new data from remote sensing, together with new methods such as geophysical tomography, will enable the development of a new generation of hydrologic/ecological theory and methods, thereby providing predictive capabilities that do not exist today. Such developments will require the integration of field measurements and experiments with atmospheric, surface, and subsurface satellite imagery. This information can be incorporated into predictive models being developed for land-use change (see Grand Challenge 7 on land-use dynamics) to increase the models' accuracy and usefulness for decision making.

Important Areas for Research

1. Improve understanding of hydrologic and geomorphic responses to precipitation. New biophysical theories and models needed to utilize the new high-resolution radar data are not yet in place. Comprehensive theories of flooding and new methods of flood forecasting would soon become possible if scientific advances enabled hydrologists and geomorphologists to take advantage of satellite images of the atmosphere and the Earth's surface. For example, the sparse network of modern semiautomated rain gauges does not capture such essential features of storms as their spatial extent and patterns of temporal intensity. With large-scale, high-resolution radar coverage and experimentally determined relationships of radar data to local precipitation, new models of the hydrologic response to precipitation could be developed to enhance forecasts of floods and their potential impacts on human settlements. Advances in understanding rainfall-runoff processes at climatic extremes would also be possible with remotely sensed data. These data could be used in combination with field measurements to construct improved maps of land cover and surface topography, and to make better estimates of soil hydraulic properties and channel dynamics.

2. Improve understanding of surface water generation and transport. Research is required to extract critical environmental-sensitivity information from satellite imagery and field instrumentation. New methods are needed to develop standard environmental indicators for surface water that can take advantage of the high resolution of precipitation forecasts. Such indicators could be used to inform and constrain process-based models of river flow and lake circulation. For example, satellite data could be used to detect contamination events and changes in water temperature, and to develop quantitative descriptions of hydrologic transport processes in rivers and lakes. Forecasting based on hydrologic and geomorphic simulations and real-time data analysis could also provide an early warning of waterborne disease outbreaks, of impending fish kills (as high water temperature indicates low dissolved oxygen content), and environmental disasters resulting from hot-water or contaminant discharges. Spatially explicit models of water and sediment distribution and movement would provide the foundation for predicting effects on aquatic organisms, including riparian species.

3. Examine environmental stresses on aquatic ecosystems. Future remote sensing capability will enable ecologists to quantify the effects of altered hydrologic regimes (for instance, from irrigation and dams) and of environmental stresses (such as pollution, erosion, and salination) on the fundamental ecological properties of aquatic systems such as biodiversity, community dynamics, primary and secondary productivity, elemental cycling, and resistance/resilience to disturbance. Such increased understanding would allow the development of creative strategies for assessing the tradeoffs between preservation and restoration of aquatic resources and demand for water.

4. *Explain the relationships between landscape change and sediment fluxes.* Future hydrologic research should be aimed at developing new concepts and quantitative physical models of sediment transport, erosion, and deposition that are based on precise topographic data of entire watersheds and high-resolution radar imagery. With improved theories of landscape evolution over a range of time scales, quantitative hydrologic and mass-transport models could become tools for anticipating environmental hazards that are the consequence of active surficial processes. Such research could help provide improved real-time warnings of landslides and mudslides; estimates of the long-term impacts of sedimentation and erosion on river morphology and consequently on navigability and flooding potential; and, when combined with analysis of land-use dynamics (Grand Challenge 7), estimates of the cumulative impacts of forest clearcutting, urban development, and other land-cover changes on water quality and on habitat as a result of changes in flooding patterns and frequencies. In addition, satellite radar data could be used to detect small changes in land-surface elevation and monitor land subsidence over vast regions due to groundwater extraction.

5. *Improve understanding of subsurface transport.* New high-resolution geophysical techniques will enable scientists to "see through" the Earth and develop a clearer understanding of the structure and behavior of subsurface water-bearing and -transmitting reservoirs (National Research Council 2000b). This understanding is beyond the reach of traditional invasive measurement methods involving well drilling and trenching. Subsurface reservoirs supply much of the nation's public water supplies, and yet many are threatened by overuse and by contamination with industrial solvents, metals, fertilizers, pesticides, and herbicides. Zones of contamination are of undetermined extent, and the migration path is often unknown. The rapidly advancing field of geophysical tomography could, for the first time, make it possible for geological scientists to observe the shallow subsurface. This type of data, combined with hydraulic information, could yield a new understanding of subsurface properties and the distribution of relative flow paths and flow barriers. The resulting hydrogeological theories and models could be used to assess declining water levels, locate subsurface contaminants, track contaminant migration, and improve the knowledge base for decisions on managing aquifers.

6. *Map groundwater recharge and discharge vulnerability.* New remote mapping capability using radar and infrared satellite data could be coupled with field measurements and new theories in hydrologic science to understand the signature of recharge areas and estimate evapotranspiration rates over vast regions. There are two critical environmental problems to be addressed. First, maintaining groundwater supplies depends on identifying groundwater recharge

areas and assessing which of these areas are threatened by depletion or contamination resulting from human activities. Second, identifying regions experiencing environmental stress due to a lack of soil moisture is key to managing agricultural production potential and assessing vulnerable aquatic habitats. New hydrologic models would make it possible to interpret high-resolution radar and infrared satellite imagery collected over time to identify and quantitatively assess impacts to recharge and discharge areas.

The hydrologic cycle is a ubiquitous part of the Earth's environmental system, so it is not surprising that this grand challenge overlaps in substantive ways with several others identified in this report. As noted earlier, land-use changes can have significant hydrologic impacts, and thus the observations and modeling efforts described here must be closely linked with those related to land use (Grand Challenge 7). The ability to predict climate variability and extreme weather events (Grand Challenge 3) is obviously a central facet of hydrologic forecasting. Biogeochemical cycles (Grand Challenge 1) are related as well, because one must understand the hydrologic characteristics of a region to estimate such phenomena as the transport of nutrients through agricultural runoff, river discharge rates, and sediment flows. Finally, the social science research discussed under Grand Challenge 6, Institutions and Resource Use, is highly relevant because of the need to strengthen institutions for water resource management, to understand the factors that drive human appropriation of water resources, and to determine how hydrologic forecasts can be used most effectively.

GRAND CHALLENGE 5: INFECTIOUS DISEASE AND THE ENVIRONMENT

The challenge is to understand the ecological and evolutionary aspects of infectious diseases; to develop an understanding of the interactions among pathogens, hosts/receptors, and the environment; and thus to make it possible to prevent changes in the infectivity and virulence of organisms that threaten plant, animal, and human health at the population level.

Practical Importance

There is a critical imperative to understand and prevent outbreaks of infectious disease in valued species, including our own. Toxic organisms and pathogens, including protists, algae, microbes, parasites, and viruses, are responsible for a major burden of disease and premature mortality among plant, animal, and human populations. The impact of natural toxins and pathogens on host populations is governed largely by factors regulating the growth of these organisms and their vectors, as well as their distribution, mechanisms of transmission/exposure (how hosts encounter pathogens), infectivity (how pathogens colonize hosts), virulence and toxicity (severity of disease), and host resistance (both host response and communicability to other hosts). These factors involve fundamental ecological and evolutionary processes, and are attributes of relationships between disease organisms and their environments, including the internal environment of their hosts. Pursuing this grand challenge would bring ecological and evolutionary understanding to bear on the problems of disease prevention and control.

Despite recognition of the importance of environmental conditions to microbial and pathogen ecology, a holistic understanding of the role of the environment in the distribution, infectivity, and virulence of pathogens remains in its infancy. There has recently been a great deal of interest in evaluating the effects of local and global climate conditions on distributions of vectors (Lindsay and Birley 1996, Martens et al. 1995, Doggett et al. 1999); however, other regional and local-scale factors may be equally important. The unanticipated effect of the Aswan High Dam on the distribution of the schistosomiasis vector is one well-known example (Abdel-Wahab et al. 1979). Uncontrolled dispersal of animal wastes can lead to harmful algal blooms in estuarine environments (Harvell et al. 1999, Fleming et al. 1999, National Research Council 2000a). Changes in the production of livestock feed may have contributed to the transfer of neurodegenerative diseases, such as mad cow disease, across species (Scott et al. 1999). And overuse of antibiotics (including use in factory farming of chickens and hogs in the United States, as well as medical practices and consumer misuse) results in the selective growth of antibiotic-resistant microorganisms (Tollefson et al. 1997, Wegener et al. 1999). The widespread use of genetically engineered crops has the potential to have a similar effect on pathogens (i.e., to select for resistance to the anti-infective agents in the crops).

Little attention has been given to the potentially important effects of environmental modification on host response. For example, exposure to ultraviolet light B (UVB) is known to inhibit immune function in humans (Morison 1989), while exposure to immunotoxic chemicals, such as polychlorinated biphenyls (PCBs) and dioxins, has been suggested as a contributing factor in the deaths of marine mammals in the north Atlantic (Ross et al. 1996). Understanding of such ecological factors in pathogen-host relationships is likely to lead to new insights about the causes of disease and new possibilities for prevention.

In meeting this challenge, a community of currently disparate disciplines must come together with the common goal of understanding the interactions between the environment and disease-causing organisms. Such research would lead to a more complete mechanistic understanding of the environmental factors altering the evolution of hosts and disease organisms, thus improving understanding of the mechanisms of infectious disease at the molecular and population levels. This improved understanding would in turn assist in the development of biological, social, and environmental controls for containing the spread of pathogens and toxic organisms; lead to guidelines for avoiding actions that encourage the development of resistance in pathogens; and help identify possible trigger events, conditions, or underlying processes that foster changes in the population dynamics and biology of pathogens and toxic organisms.

Scientific Importance

During the past several decades, it was commonly believed that, at least in the developed countries, disease pathogens had been permanently surpassed in public health importance by noninfectious chronic diseases of aging. During this period, research on the role of pathogen/toxin exposures in human disease suffered relative neglect. Recently, however, the issue has attracted renewed scientific concern for several reasons. First, pathogens are now recognized to play a causal role in many chronic diseases and conditions, including cardiovascular disease, neuropsychiatric disorders, infertility, and ulcers. In addition, infections such as tuberculosis, malaria, and pneumonia have reemerged, and newly recognized pathogens—HIV, Nipah virus, West Nile virus, Lyme disease, transmissible spongiform encephalopathies, the hepatitis viruses—have grown in medical importance. We do not fully understand how or why these pathogens episodically present public health threats to humans, birds, and other animals. Disease-related pathogens such as Epstein-Barr virus and the tuberculosis bacillus are found in large numbers of clinically healthy individuals, suggesting that established models linking exposure and transmission to illness need amplification to incorporate virulence and changes in host susceptibility (Morris and Potter 1997).

Both human and other populations are affected by changes in pathogen distribution and virulence, but the mechanisms by which this occurs are not yet well understood. Often, as is the case with *Pfiesteria*-associated fish kills on the East Coast of the United States (Silbergeld et al. 2000a), domoic-acid-induced deaths of sea lions in California (Scholin et al. 2000), and Nipah virus in Malaysia (Chua et al. 1999), nonhuman species are the first sentinels of change. Deaths of crows and other birds provided the key to identifying West Nile Virus in New York in 1999 (Lanciotti et al. 1999). Similarly, understanding of the zoonotic (i.e., animal-related) aspects of immunodeficiency virus infection has importance for human health (Hahn et al. 2000). Given these linkages, recent reports of the role of parasitic infections and environmental stressors in causing amphibian deformities may reflect a sentinel event that will eventually be detected in other species (Burkhart et al. 2000).

If campaigns to reduce or eliminate major diseases such as tuberculosis and malaria are to be successful, the research community will need to design global interventions and monitor the efficacy of these investments. The phenomenon of chemotherapeutic and antibiotic resistance in many pathogens suggests that ecological approaches to disease control may be a necessary supplement to new drugs and vaccines (Morse 1993). As part of these efforts, it is important to anticipate the impacts of environmental change on disease prevalence. At present, however, little is being done to evaluate the impacts of small- and large-scale environmental perturbations on host-pathogen-toxin relationships. Projections of future global change indicate that we may see the migration of new insect vectors capable of transmitting diseases such as malaria into previously uninhabitable geographic regions (Martens et al. 1995). Such changes in ecological dynamics will require both anticipation and adaptive responses by societies living in affected regions.

Because of their rapid growth rate and large populations, microbial pathogens can evolve very quickly, and these evolutionary mechanisms allow them to adapt to new hosts, produce new toxins, and bypass immune responses. Many of the weapons used against microbes (drugs, vaccines, pesticides) can inadvertently contribute to the selection of adaptations that enable pathogens to proliferate or nonpathogens to acquire virulence. This evolutionary perspective, sometimes referred to as "Darwinian medicine" (Ewald 1996, Williams and Nesse 1991), can greatly improve our understanding of pathogen behavior and host response, as well as our ability to design appropriate intervention and disease treatment strategies.

While the focus of this challenge is on the biological and ecological understanding of infectious diseases, it is important to recognize that social, behavioral, and economic factors also play a role in transmission, infection, and disease among both human and animal populations (Kiesecker et al. 1999). Changing patterns of housing in the United States have facilitated the transmission of Lyme disease via ticks to humans. Likewise, cultural change and lifestyle choices are involved in the transmission of the HIV virus and the pathogens responsible for other sexually transmitted diseases. Economic choices in animal husbandry have driven the increasing use of antibiotics to promote rapid growth, and alterations in meat processing led to the emergence of mad cow disease (DuPont and Steele 1987). These examples illustrate the range of social issues that need to be considered to fully address the population health impact of infectious diseases.

To make progress in understanding emerging infections,[1] it is necessary to develop an ecological understanding of disease. Developing such an understanding requires in turn the integration of research concepts from theoretical ecology, immunology, genetics, evolution, population biology, and the environmental and social sciences. To quote from Wilson (1999, pp. 308-309):

---

[1] We define emerging infections as those whose incidence has increased within the past two decades or whose incidence threatens to increase in the near future as a result of the spread of a new agent, the recognition of a previously undetected infection in a population, the realization that an established disease has an infectious origin, or the appearance of a known infection after a decline in incidence (National Research Council 1992).

Studies of emerging infections typically rely on disease, organismic or syndromic approaches. By contrast, understanding the process of disease emergence involves studying the origins and ecology of emerging infections. . . . Tools used to study and understand disease emergence include mathematical modeling, geographic information systems, remote sensing, molecular methods to study the genetic relatedness of organisms, and molecular phylogeny. Paleobiology, paleoecology, and studies that allow the reconstruction of past events may help inform future research and policy. The study of disease emergence must be at the systems level and must look at ecosystems, evolutionary biology, and populations of parasites and hosts, whatever their species.

Scientific Readiness

Four advances make this challenge appropriate for strategic investment at this time.

The ability to sequence the genomes of pathogens, parasites, and vectors. The technology needed to sequence the entire genomes of selected pathogenic organisms and vectors now exists. As the sequences become publicly available, it will become possible to test hypotheses related to resistance, virulence, and adaptation.

Improved understanding of gene-environment interactions in host immune response. Enhanced understanding and improved methods for acquiring further knowledge of the molecular determinants of host immune response (including but not limited to genetics) allow for more sophisticated epidemiological and zoonotic surveillance of host resistance at the molecular level than was previously possible.

Increased computing power and developments in theoretical population biology. New techniques and capacity for nonlinear dynamic modeling allow for the development and testing of more complex models that integrate information from the genome to the ecosystem. These models incorporate new insights from theoretical population genetics, evolutionary biology, and population ecology.

Data acquisition systems for ecosystem monitoring. As discussed in relation to Grand Challenge 4, Hydrologic Forecasting, and Grand Challenge 7, Land-Use Dynamics, new systems for monitoring (including satellite remote sensing) and for recording data (e.g., geographic information systems) have the potential to provide ecosystem-level information that can be incorporated into the above models (Lobitz et al. 2000, Hay et al. 1998). These remote systems can now be linked to molecular biomonitoring systems to anticipate changes in pathogens or toxin distribution (Rhodes et al. 1998). These methods and the associated predictive models are ready to be validated by epidemiological surveillance.

Important Areas for Research

1. Examine the effects of environmental changes as selection agents on pathogen virulence and host resistance. Pathogens, vectors, and hosts are affected by local and global changes in the environment, including physical, chemical, and climatic alterations, as well as direct human influence. Vectors and pathogens tend to adapt to environmental change through exploitation of advantageous ecosystems and through evolutionary selection that favors survival in an altered environment, which can be accompanied by the exchange of favorable genes within and across species. Antibiotics are part of the chemical environment of pathogenic organisms, and microbial resistance to antibiotics is a growing problem in managing disease (National Research Council 1999g). The ecology and molecular biology of drug resistance is understudied, and there is a need for more accurate and complex models that incorporate the mechanisms by which organisms acquire and shed resistance, the phenomenon of polyresistance, and gene transfer across organisms and populations. The ecology of pathogens is also affected by human settlement patterns, as well as agricultural, sanitation, and development practices. Research is needed to improve understanding of how pathogens adapt to all of these selection agents.

Chemicals can also act as environmental stressors that affect pathogen virulence and immune response. Examples are the demonstrated loss of immunity to malaria brought on by low-level exposures to mercury (Silbergeld et al. 2000b) and the increased susceptibility to bacterial respiratory infections caused by exposure to air pollutants such as ozone. In general, these relationships are poorly understood, and they represent an avenue of research ripe for important discovery and offering opportunities to test existing models.

2. Explore the impacts of environmental change and variability on disease etiology, vectors, and toxic organisms. Changes in climate, land use, water quality, natural species distribution, and species introduction brought on by human activity have the potential to alter the spread and impact of pathogens, parasites, and toxic organisms. Scientists have limited knowledge of the ecological variables that promote or deter the rapid growth of toxic organisms, such as algal blooms, or the environmental conditions that may elicit the production of natural toxins. Likewise, we have little specific understanding of how predicted widespread climate and land-use changes, changes in water and waste management, alterations in the biogeochemical cycles of nutrient compounds, or changes in food production systems may affect the ecology and spread of disease organisms on small or medium scales. Diseases such as malaria, dengue, and cholera may be especially sensitive to environmental and climate change (Lindsay and Birley 1996, Patz et al. 1998, Colwell 1996). The introduction of bioengineered organisms may also alter the ecology of pathogens, vectors, and hosts by disturbing the ecosystems of pathogen-transmitting and predatory organisms. New experimental and modeling approaches are needed to help in predicting how environmental changes will interact with pathogens and toxic organisms and in developing a sufficient understanding of the mechanisms of these interactions to allow the mitigation of potential harmful effects.

3. Develop approaches to surveillance and monitoring. There is an urgent need to develop new integrative approaches for monitoring sensitive indicators of change that will enable anticipation and mitigation of infectious diseases in humans and other species. Current methods for tracking changes in host-pathogen ecology depend on the detection of infections in target populations. Recent studies associating short-term climate change events with outbreaks of Rift Valley fever in Africa (Shimshony 1999, Linthicum et al. 1999), cholera in Peru (Franco et al. 1997), malaria in Kenya (Hay et al. 1998), and Hanta virus in the United States (Morse 1993), are highly informative in this respect. This type of monitoring can be improved by establishing disease registries that permit molecular identification of new diseases or new variants of existing diseases. Recent experience in New York City, where, as noted, West Nile virus went undetected until bird deaths were discovered, points to the importance of sophisticated methods of surveillance in multiple populations (Lanciotti et al. 1999) and to the urgent need for disease registries at the international level, given the opportunities for transboundary movement of pathogens and infected hosts (Roeder et al. 1999). Currently, it is very difficult to obtain a comprehensive picture of global infectious disease trends since in many parts of the world, basic epidemiological data are not collected or are not shared for political reasons.

Existing programs must be expanded to include surveillance of the population ecology of zoonotic hosts, pathogens, vectors, and toxic organisms. Conducting such surveillance will necessitate developing and monitoring molecular and genetic markers of change in disease organisms, and on using geographic information systems to incorporate ecological data from remote and in situ observations with geographically explicit data on the populations of pathogens and toxic organisms. New methods developed to forecast blooms of toxic algae, incorporating both remote and on-site monitoring of population dynamics and toxin production (Rhodes et al. 1998), can be applied to other surveillance systems and theoretical models of outbreak.

4. Improve theoretical models of host-pathogen ecology. The contributions of theoretical ecology and population biology must be incorporated into biomedical research on the prediction of infection and disease through the development of complex models of host-pathogen ecology capable of predicting infection, transmission, and disease incidence. The capacity for such cross-disciplinary endeavors needs strengthening, beginning with enhanced and redesigned training and research. In addition, complex, interdisciplinary prospective experiments must be explicitly designed to test hypotheses derived from the models. The funding of this research will require an unprecedented level of cooperation among granting agencies across the relevant basic and clinical disciplines, as well as increased support for international collaboration in research and surveillance.

GRAND CHALLENGE 6: INSTITUTIONS AND RESOURCE USE

The challenge is to develop a systematic understanding of the role of institutions—markets, hierarchies, legal structures, regulatory arrangements, international conventions, and other formal and informal sets of rules—in shaping systems for natural resource use, extraction, waste disposal, and other environmentally important activities.

Practical Importance

Most human uses of natural resources and impacts on environments are mediated by rules and regulations—from village-based land tenure systems to international accords to regulate the release of CFCs to the atmosphere—related to the resources' provision, access, and use. These sets of rules and regulations are called institutions. For most of history, such institutions evolved locally in accordance with intimate associations between resources or environments and their human uses. Recently, however, such institutions have increasingly been designed by state or extra-state entities to address large-scale, even global, problems of open-access resources or environments (e.g., those with no enforceable rules regarding their use, such as many open-ocean fisheries). Institutions may act to limit demands on resources or to generate additional demands. In either case, understanding the character and role of institutions is pivotal to understanding human-environment interactions and to assessing the potential consequences of the many institutions emerging at multiple scales to deal with environmental change.

The range of institutions regulating access to and use of land, water, minerals, the atmosphere, forests, fisheries, and other natural resources is as broad as the range of their impacts. For example, many thousands of water management institutions—some 20,000 governing units in the United States alone—provide rules for water rights, each having different impacts on entitlements to water and on water resources. These institutions also establish a variety of rules for paying for water use. The water-use rules established by institutions can have widely varying effects:

• Property institutions that give individuals the right to pump the Ogallala aquifer of the High Great Plains of the United States have led to dramatic declines in this source of fossil water while increasing grain production for America and much of the world.

• In contrast, communally based regulation of irrigation systems in the Philippines has limited water withdrawals and provided for crop requirements over long periods (Siy 1982); however, they produce little beyond immediate consumption needs.

• A diffuse system with multiple institutional controls led to significant ecological change in Lake Erie.

• The largest catastrophe to any major water ecosystem, the destruction of the Aral Sea ecosystem in central Asia due to the sea's drying up, followed from the actions of command-and-control water institutions that fostered excessive water withdrawal and contamination (Glazovsky 1995).

• A polycentric system involving private associations, multiple city and county governments, the state-level court system, and special districts facilitated the reversal of a severe overdraft of coastal groundwater basins in Southern California that supported a growing urban economy (Blomquist 1992).

Management of water resources can benefit from improvements in hydrologic forecasting outlined under Grand Challenge 4. Better hydrologic forecasts alone, however, are not sufficient to inform the design of effective water management institutions.

Scientific Importance

The above examples illustrate that resource use is mediated or determined by institutions and is affected, often in major ways, by the structure and efficacy of the institutions. They also illustrate that no single institutional form is best for all resources or all situations. What we do not yet know is the conditions under which each institutional type works well or the factors that determine the environmental and social consequences of different institutional forms. The full range of institutions controlling critical resources and environments worldwide is not well documented. The fundamental characteristics and attributes of these institutions have not been examined comparatively and with the aim of clarifying how different institutions work under differing sets of human-environment conditions (e.g., rapid technological change, climatic variation, increases in resource demand).

The general lessons that can be gleaned are illustrated by various work conducted during the past decade. For instance, open-access resource systems that face increased demand are subject to rapid extraction that threatens ecosystem functioning and human welfare (see, e.g., Bromley 1992, Kasperson et al. 1995). This pattern tends to be reproduced when local institutions that enforce rules for resource use are challenged, corrupted, or destroyed and are either not replaced or replaced by externally constructed institutions, as has often happened when colonial powers or central governments have assumed responsibility for resource management. Examples include deforestation in southeastern Asia (Agrawal 1999) and wildlife management in Africa (Gibson 1999). There has been considerable relevant theoretical work on the design of markets (e.g., Baumol and Oates 1988, Loehman and Kilgour 1998), and there is a growing body of empirical work on common-pool resource institutions (e.g., Ostrom 1990);

however, understanding of the inner workings of various classes of institutions is still in its infancy.

Although resource institutions may need to change in response to rapidly changing environmental conditions, little is known about the characteristics of institutions that predispose some of them to adapt successfully. Some private-ownership systems (e.g., those associated with technologically vital metals) or open-access systems (e.g., some fisheries) do not provide effective incentives for the conservation of environmental goods (e.g., National Research Council 1999d,e). Others fail to provide the accurate information about biological and economic processes that is needed to adjust to change (Moxness 1998). The overarching scientific challenge is to develop a sufficient understanding of different institutions and their responses to change so that institutional design choices can be based on empirically grounded knowledge, not just intuition. For example, are institutions more adaptive and their resource bases better protected if they encourage small-scale experimentation, collect accurate performance data, and seek to monitor biophysical feedbacks and surprises?

Scientific Readiness

During the past several decades, theoretical and empirical advances in social science have significantly increased the capacity to address resource and environmental management institutions in a systematic fashion and to understand the environmental and social consequences of different institutional forms. The field stands at the threshold of substantial progress as a result of new multidisciplinary empirical studies of resource institutions; advances in institutional design theory in economics and political science; and developments in institutional, environmental, and resource economics.

An interdisciplinary research community has matured. A shared set of analytical concepts has been developed and applied by researchers in several relevant disciplines. Communication among researchers occurs in an international scientific society and through a new international research project under the International Human Dimensions Programme (IHDP) on Global Environmental Change. This project has established research foci on the role of institutions in causing and confronting global environmental changes, the factors that distinguish successful from unsuccessful institutions, and the prospects for redesigning institutions to confront environmental challenges (Young 1999).

A large body of case material has been gathered and organized around key concepts (Hess 1999). Systematic research enables scholars to identify who is eligible to use and harvest a resource at what quantity, location, and temporal order; the technology that can be used for harvesting; how provision and maintenance activities are organized; how decisions about resource management are made; what kinds of information are provided; and what outcomes are achieved in terms of economic returns, accountability, and sustainability. Further, attributes of resources and their users that affect the costs and benefits of organizing resource regimes have been identified. Ongoing field research is now beginning to test the hypotheses derived from this growing body of theory (Gibson et al. 2000).

Studies drawing on the theory of markets have identified and analyzed several promising new institutional approaches for dealing with resources that normally have no market value and are consequently subject to overexploitation. A prominent example is the construction of a market for pollution credits, first proposed in the late 1960s on the basis of economic analysis (e.g., Crocker 1966, Dales 1968) and implemented in the United States in the 1980s by creating a tradable property right to emit sulfur dioxide into the atmosphere. The experiment succeeded beyond expectations (Stavins 1998), and further applications to CO_2 emissions is being considered. Similarly, individual transferable quotas (ITQs) have sometimes been implemented successfully to control fishing (National Research Council 1999d,e). Enough practical experience is now being gained to permit systematic evaluation of the empirical performance of these new institutions. Adaptive management systems have been shown to increase the resilience of complex environmental systems (Berkes and Folke 1998). Other promising institutional innovations involve local-national comanagement of resources (Keohane and Ostrom 1995). Although science has not yet specified the range of conditions that favor successful implementation of each such institutional form, it is now possible to state clear hypotheses and evaluate them empirically.

Research on institutions is incorporating the biological and physical sciences of environmental systems. Social scientists are beginning to work with natural scientists to develop more effective models of how human actions and institutions interact with the environment (National Research Council 1998, Ostrom et al. 1999). On the technological front, recent advances in remote sensing are making it possible to monitor many resources in standard ways across space and time, providing new ways to measure the effects of different resource management practices (National Research Council 1998).

Important Areas for Research

1. Document the institutions governing critical lands, resources, and environments. Various environmental studies, especially those requiring models to project the impacts of change, need information on the key institutions governing the land, resource, or environmental problem of concern. Thus, research on topics ranging from land-use change, to fishing stocks, to freshwater resources, to atmospheric dynamics ultimately requires consideration of the controlling institutions, especially for regional and global models. These institutions encompass national laws and regulations; market structures; property rights systems; and informal practices governing resource access, use, and exchange. Likewise, research intended to be useful for planning and design purposes requires an understanding of existing resource institutions and the incentives they create for resource users.

2. Identify the performance attributes of the full range of institutions governing resources and environments worldwide, from local to global levels. Institutions, whether they evolved over the long run to govern a specific, local resource or were recently designed to reduce damage to a global system (e.g., ozone depletion), have certain attributes that function to achieve the goals of the systems they govern. These attributes have not been addressed systematically for all forms of institutions, and their performance under different sets of conditions has not been assessed. For example, various common property systems have served well to conserve resources or environments over long periods during which demands on the systems were relatively low and static in quality (e.g., Netting 1981, Ruttan 1998). While some of these systems have adapted well to major changes, others have not served well under conditions of high resource demand, major changes in resource-extracting technology, or rapid changes in social and political conditions. In contrast, privately owned institutions may serve the economic interests of the individual user and afford an opportunity for adjustments in resource use as conditions change, but they may be poorly suited to handling environmental problems arising from landscape-level functions, such as loss of biodiversity following from landscape fragmentation.

Research on performance attributes should address both traditional institutional forms and new forms, such as those that attempt to manage resources by creating markets for emission or extraction rights. The research should also address both the intended purposes of institutions and their unintended consequences, including effects on resources other than those they are intended to manage. An important performance attribute of resource management institutions is the way they incorporate information about resources from both local observers and organized science. Especially where resources are under threat, successful resource management is likely to depend on institutions' ability to entrain decision-relevant science and to use its outputs in a timely manner.

3. Improve understanding of change in resource institutions. Most resource institutions evolve over the long-run in response to changes in their resource bases and their social and economic contexts. For example, broad shifts of power between national and local governments and of influence between governments and transnational organizations (e.g., corporations, intergovernmental and nongovernmental organizations) can create a change in resource institutions. Although external forces in the environment or in government may press local institutions to change, institutional changes are usually contested by interested and vested parties. In some cases, institutions are predisposed to making some kinds of changes but not others. A major scientific need is to understand the conditions both within and external to institutions that affect their patterns of adaptive change.

4. Conceptualize and assess the effects of institutions for managing global commons. Much attention is now being given to the design of new institutional forms for controlling previously unregulated global common-pool resources and environmental conditions, such as stratospheric ozone depletion, atmospheric CO_2, and oceanic dumping of waste materials. Both global agreements and national implementation are required. Research on this topic should focus on the effects on resource use of different combinations of policy instruments and monitoring activities, and on the effects of differences and conflicts among the incentive structures of local, national, and global institutional arrangements.

GRAND CHALLENGE 7: LAND-USE DYNAMICS

The challenge is to develop a systematic understanding of changes in land uses and land covers that are critical to biogeochemical cycling, ecosystem functioning and services, and human welfare.

Practical Importance

Humans have dramatically altered the Earth's surface. These changes in land cover—the land surface and immediate subsurface, including biota, topography, surface water and groundwater, and human structures—are so large and rapid that they constitute an abrupt shift in the human-environment condition, surpassing the impacts of all past epoch-level events (e.g., the domestication of biota, the industrial revolution) since the rise of the human species. Indeed, they approach in magnitude the land-cover transformations that have occurred at transitions from glacial to interglacial climate (Meyer and Turner 1994, Ramankutty and Foley 1999).

Human-induced land-cover change to date, especially tropical deforestation, has been a primary influence on global atmospheric circulation patterns and a major contributor to observed increases in atmospheric concentrations of CO_2 (e.g., Houghton 1994, International Geosphere-Biosphere Programme 1999). The annual rate of tropical deforestation remains high, hovering near 1.0 percent during the 1980s (Tolba and El-Kholy 1992). Human use of land, that is, what people do to exploit the land cover, has been the primary culprit in the estimated 2.95 million km^2 of soils whose biotic function has been significantly disrupted by chemical and physical degradation—including 1.13 million km^2 disrupted by deforestation and 0.75 million km^2 by grazing. In addition, agriculture currently consumes 70 percent of total freshwater used by humankind, much of which is accounted for by the rapid expansion of irrigation, which annually withdraws some 2,000-2,500 km^3 of water.

These and other human-induced changes are major contributors to global climate change, to the loss of global biotic diversity, and to the reduced functioning of ecosystems and the essential services they provide to humans. Land-use change continues to contribute significantly to anthropogenic releases of CO_2 to the atmosphere, changes in hydrologic dynamics and nitrogen cycling, and alterations in habitat for almost all terrestrial species. Land-use changes can also interfere with the migration of some species and facilitate the spread of disease vectors (Meyer and Turner 1992). And through their impacts on ecological services, land-use and land-cover changes affect the ability of biological systems to yield enough food, fiber, and fuel to meet human needs (Vitousek et al. 1997a).

Thus, land-use and land-cover dynamics and their spatial patterns play a significant role not only as drivers of environmental change, but also as factors increasing the vulnerability of places and people to environmental perturbations of all kinds. Improved information on and understanding of land-use and land-cover dynamics are therefore essential for society to respond effectively to environmental changes and to manage human impacts on environmental systems.

Scientific Importance

The basis for a science of land-use dynamics is beginning to emerge (e.g., Skole and Tucker 1993). However, regional and global-level stocks of most land covers and uses, including such essential categories as forest and grassland cover, agricultural uses, and urban and suburban settlement, are still poorly documented and monitored. Theory and assessment models used to address land dynamics are mainly static, economic sector-based, and nonspatial, and do not account for neighboring uses; the roles of institutions that manage land and resources; or biophysical changes and feedbacks in land use and cover, including climate change and anthropogenic changes in terrestrial ecosystems. Such inadequacies must be redressed if we are to achieve a robust understanding of these phenomena and provide the kinds of projections required to conduct environmental planning and to ensure the sustainability of critical ecosystem functions. In particular, it is necessary to improve understanding of which land units change, how, where, and why.

A growing interdisciplinary research community stands poised to document, develop theory, and provide robust regional models of land-use/cover change. Research efforts are under way worldwide to address almost all land covers and uses. Certain types of changes have been identified as especially critical and should be the focus of immediate concern: deforestation and its opposite, afforestation; pasture creation; grassland degradation; intensification of agriculture; and urban-industrial spread, including suburbanization. Of the first four, three types of change focus on the spatial magnitude of terrestrial land covers, while the intensification of agriculture deals primarily with increased water and chemical inputs to cultivation. Urban-industrial spread is important even though it involves only a small percentage of the total land surface under human management; for example, from 1982 to 1992, a relatively modest 25,800 km^2 of agricultural land in the United States was converted to urban or built-up uses (Vesterby et al. 1997). The changed parcels, however, often constitute prime lands for cultivation with concomitant cropping infrastructures, as in the case of the spread of megacity complexes worldwide. Urban development affects hydrologic processes as well (e.g., effects of paving on runoff and of urban heat islands on storms).

Close inspection by the research community has begun to illuminate the nuances of land-cover dynamics and to challenge the conventional wisdom on a number of fronts. For example, studies of deforestation in Amazonia reveal that as much as 31 percent of formerly cut forest is in various stages of regrowth (Alves and Skole 1996), with significant implications for estimates of carbon emissions and of annual rates of change in the forested areas of the tropical world. Similarly, studies of land changes in the humid savannas of West Africa indicate that woody biomass has been increasing and continues to do so in areas claimed by some observers to be experiencing desertification (Bassett and Koli 2000). Inventories of ecosystems in the United States during the 1980s demonstrate an accumulation of carbon, largely through afforestation, equivalent to between 10 and 30 percent of U.S. fossil fuel emissions (Houghton et al. 1999). And changes in land use and cover affect local and regional climates; in South Florida, for instance, a drier, warmer interior during the months of July and August has followed the expansion of agriculture (Pielke et al. 1999).

Documentation and monitoring of these and other trends provide an observational base for efforts to improve understanding of the dynamics of land change, projections of climate change (by better specifying the contribution of land cover), and estimates of the full range of impacts of various land-cover "swaps" intended to reduce CO_2 emissions (e.g., trading energy units from power plants in temperate industrialized countries for afforestation in the tropics). The international, interdisciplinary research community has begun to address the explanatory power of relative location (the effects of surrounding land uses on the potential for a unit of land to change), path dependency (the role of previous conditions and trajectories of change in constraining options for future change), biophysical feedbacks (e.g., effects of nutrient depletion with cropping), land and resource institutions (e.g., land tenure), and induced innovation (the capacity of agents and society to innovate internally as conditions change). Understanding the interrelations among these factors is often key to explaining land-use change and its environmental and social effects. For example, the highest recorded emissions of the greenhouse gas nitrous oxide and the ozone-affecting gas nitric oxide from soils have been linked to policy-influenced cropping procedures in northern Mexico's irrigated "wheatbasket" (Matson et al. 1988). Likewise, different tenure institutions controlling land uses and stocking strategies in Rajasthan, India, have led to significant differences in grassland quality and presence of trees (Robbins 1998).

Researchers are also beginning to demonstrate the value of spatially explicit analytical approaches as compared with nonspatial measures of the magnitude of change (Lambin 1994, Turner 1990). For example, by including the spatial heterogeneity of the landscape and modeling interactions between land users and other decision makers, recent economic models of suburbanization in the Patuxent watershed of Maryland have improved the explanation of land use and its change over what could be achieved with traditional nonspatial and noninteractive models (Bockstael 1996).

As a result of such advances, the research community is now poised to develop at least four types of spatially explicit, integrative, explanatory land-change models: (a) those based on behavioral and/or structural theory linked to specific geographic locations, (b) those drawn from changes registered in remotely sensed imagery, (c) hybrids of these two types, and (d) dynamic spatial simulations (DSSs) that offer projections under different sets of assumptions (Frederick and Rosenberg 1994, Liverman et al. 1998). Theory- and imagery-based models are used to explore explanations of change and to provide near-term (5-10 years) projections under differing sets of assumptions. They permit tests of the applicability of various theories for different areas and conditions and the coupling of local-, regional-, and global-scale models by land cover or use type. An example is the fit of the Yucatan Peninsula to local versus pantropical models of tropical deforestation. DSSs, on the other hand, address scenarios over the longer term (more than 10 years) by making the agents, structures, and environment interactive and dynamic. For example, a DSS can examine how changes in the structures governing land access change agents' decisions about use, and in turn, the environmental qualities of the land feed back to agents and institutions governing land access.

Scientific Readiness

In addressing this challenge, new research would characterize regional variations in the pace, spatial scale, and magnitude of change in critical land uses and covers. It would identify the ways in which individual, household, and institutional actors and structures affect these changes and, in turn, respond to their biophysical consequences. The research would also develop increasingly robust models for addressing these dynamics in spatially explicit ways at different spatial scales and in relation to multiple sectors of human activity. Several recent developments make the area ripe for further advances, promising to transform land-use/cover change science.

Improved databases on land cover and land use. Key organizations and agencies are improving their databases on land in a manner consistent with the needs of global change science. For example, the Food and Agriculture Organization is leading an effort to create an international land-use typology and to employ this typology in its country-wide compilations of land conditions. The new Landsat 7 satellite will provide frequent worldwide imagery of land cover from the Thematic Mapper system at costs affordable to the community of land researchers.

Advances in imagery analysis and geographic information science. These developments are providing the tools and analytical capacity needed to address land-use/cover dynamics spatially and to link social science and biophysical data. These capabilities and the emergence of other kinds of spatially explicit data have triggered interest in land use among new communities of researchers, such as demographers and economists (National Research Council 1998), and have inspired researchers to develop various modes of spatially explicit, multi-sectoral land-change models that begin to integrate statistical, diagnostic, and prognostic approaches at the regional level.

Advances in the analysis of spatial data. Advances are being made toward solving some major methodological problems involved in the analysis of spatial data. For example, spatial autocorrelation, the tendency for two points in close proximity on the Earth's surface to have similar properties, invalidates the use of statistical tests that assume independence among observations. Methods are being developed to account for spatial autocorrelation and explore its properties (e.g., Bailey and Gatrell 1995) and to analyze variables as functions of spatial location (e.g., Goovaerts 1997). Progress is also being made in finding ways to improve the drawing of inferences from large-area data—often the only data available—to small-scale processes (e.g., King 1997) and in understanding how the results of spatially aggregated data analysis depend on the basis of aggregation (Openshaw 1983).

Increased inter- and multidisciplinary interest in the science of land-use/cover change. Stimulated by various international and national research programs, formerly diverse sets of researchers worldwide are engaged in collaborative ventures to create integrative approaches to the study of land-use/cover change. In the United States alone, 25 such teams have been formed by NASA's Land-Cover and Land Use Change program, with strong linkages to several of the centers of excellence sponsored by NSF. This figure is substantially larger at the international level. Additionally, the U.S. Geological Survey, working with the Environmental Protection Agency, is supporting research projects on land-cover trends and on urban dynamics, and NSF sponsors a small Human-Environment Regional Observatory project. These federal initiatives are an important beginning, but still lack the coordination, scope, and focus on integrated land-change models called for under this challenge.

Important Areas for Research

1. Develop long-term, regional databases for land uses, land covers, and related social information. These databases should emphasize the critical land uses/covers of forest, grasslands, agriculture, and urban-industrial settlement and should include complementary demographic, economic, and institutional information. Work on developing useful land-cover data must include efforts to improve the accuracy and reduce the uncertainty of vegetation classification from remote observation platforms.

The research community has identified regional data observatories and archives as essential. They are, however, extremely difficult to establish and sustain, and few if any interdisciplinary exemplars exist. Increased temporal resolution of high-spatial-resolution, space-based imagery is needed, along with reduced costs of such data for individual researchers. The issue of the confidentiality of social data also requires attention.

2. Formulate spatially explicit and multisectoral land-change theory. Research in this area should address the causal roles in land dynamics of relative location, past uses (path dependency), land and resource institutions, and biophysical changes and feedbacks (e.g., climate change, nutrient depletion), and should determine the significance of regional variations in these relationships. Until now, land-change theory has been crafted in relatively simple terms and focused on specific economic or land sectors or products (e.g., agriculture or timber production). Understanding the causes and implications of land-use/cover change requires the development of theory that can account simultaneously for changes in multiple uses and covers by accounting better for the complexity of interactions that stimulate these changes. To achieve this aim, improved understanding of how agents and social structures behave or operate over space is required, along with better statistical methods that permit hypothesis testing and model validation. It is also important to understand the ecological consequences of land-use change and how ecological changes can influence land use.

3. Link land-change theory to space-based imagery. Space-based imagery offers one of the few ways to scale analysis up spatially beyond the local level. Research in this area would push the boundaries of land-cover change detection from space and develop and test imagery-led models of change that could be coupled or merged with models based on theory (actors and/or structures). The research would also press imagery analysis to detect variables or develop proxy measures important to the human science of land change. The potential of this line of research will expand as new remote platforms offer observations at increasingly finer scales, suitable for detecting human activities not previously observable from space.

4. Develop innovative applications of dynamic spatial simulation techniques. Research in this area would exploit recent gains in computing resources and techniques. It would (a) extend dynamic spatial simulation techniques to model the distinct temporal and spatial patterns of land-use and land-cover change; (b) connect these models to extant and pending theoretical frameworks that accommodate the complexity of, and relationships among, socioeconomic and environmental factors (see research area 2 above); (c) establish common validation and replication protocols necessary for determining the robustness of model outcomes under different assessment scenarios; (d) consider the value of information and the role of uncertainty in determining model outputs; and (e) examine the utility of dynamic spatial simulation models for land managers and government decision makers.

GRAND CHALLENGE 8: REINVENTING THE USE OF MATERIALS

The challenge is to develop a quantitative understanding of the global budgets and cycles of key materials[2] used by humanity and of how the life cycles of these materials may be modified. Among the materials of particular interest for this grand challenge are those with documented or potential environmental impacts, those whose long-term availability is in some question, and those with a high potential for recycling and reuse. Examples include copper, silver, and zinc (reusable metals); cadmium, mercury, and lead (hazardous metals); plastics and alloys (reusable substances); and CFCs, pesticides, and many organic solvents (environmentally hazardous substances).

Practical Importance

The extraction, use, and dissipation of technology-related materials affect humans and natural ecosystems in a myriad of important ways. First, toxic elements such as cadmium, mercury, and lead accumulating in the environment can have important negative impacts on human health (see, e.g., Thomas and Spiro 1994). An understanding of the flows of these elements and of the technological and cultural factors that drive those flows is required to mitigate these harmful effects and reduce exposure levels over the long term. Second, recovery and recycling of valuable elements such as platinum or copper can be accomplished at only 10-20 percent of the energy cost of refining these elements from natural sources (Schuckert 1997). Finally, understanding where these elements are lost during manufacturing processes and where in the environment they ultimately come to reside is necessary in considering whether to recover them.

With the changes brought about by population growth, rapidly evolving technology, more intensive agriculture, and increasing energy usage, global use of technological materials is expected to grow by as much as a factor of four

---

[2] "Materials" includes elements, compounds, alloys, and other substances created or mobilized by human activities, except it specifically excludes the elements that constitute the grand nutrient cycles—carbon, nitrogen, sulfur, and phosphorus. (The cycles of these elements have historically been dominated by natural processes, though human activities are now important perturbers; these cycles are the subject of Grand Challenge 1, Biogeochemical Cycles. Because of their association with human uses, the discussion that follows often refers to the materials of interest as "technological" or "technology-related.")

during the next several decades. The cycles of many important materials are in rapid fluctuation, with existing reservoirs changing in size and new ones being added, and timely analysis is needed to understand some of these changes. For example, we are approaching local toxicity thresholds for some materials (Environmental Protection Agency 1998), and the availability (at reasonable cost) of certain materials essential to manufacturing is becoming threatened (Kesler 1994). New compounds and other substances are constantly being incorporated into modern technology and hence into the environment, with insufficient thought being given to the implications of these actions. All of these issues assume added importance in urban areas, which concentrate flows of resources, generation of residues, and environmental impacts within spatially constrained areas. From a policy standpoint, reliable predictive models of material cycles could be invaluable in guiding decisions about issues related to fossil fuel use, energy production, agricultural practices, and a wide range of other topics relating to human-environment interactions (Allenby 1999).

This grand challenge centrally encompasses questions about societal-level consumption patterns, since consumption is the primary force driving human perturbations of material cycles. Social scientists are exploring many questions about consumption patterns that are relevant to the issue of material cycles, such as the reasons for the large variations in consumption of resources among different cultures (National Research Council 1997); the factors that drive changes in consumption patterns over time (Organization for Economic Cooperation and Development 1997); whether policy initiatives influence these patterns; and if so, which policies are most effective for any given situation. These questions relate also to Grand Challenge 6, Institutions and Resource Use.

Scientific Importance

The basic framework for understanding the flows of materials is the "budget," in which short- and long-term reservoirs are identified, and the flows between the reservoirs are quantified (e.g., Graedel and Allenby 1995). No overall pictures of generation, use, and fate have yet been produced for materials whose cycles are dominated by technology; our understanding of the budgets and cycles of nutrient compounds of carbon, nitrogen, sulfur, and phosphorus (addressed by Grand Challenge 1, Biogeochemical Cycles) is far more advanced. The construction of budgets for technological materials would be a natural outgrowth of the interaction of environmental science and the emerging discipline of industrial ecology, and would follow directly from the theoretical and analytic approaches developed for the major biogeochemical cycles (e.g., Bolin and Cook 1983). In fact, part of the scientific excitement generated by these questions is that they can be adequately addressed only through close collaboration among specialists in the natural sciences, the social sciences, and a variety of engineering disciplines to achieve the following:

- Understanding the operation of the natural cycle of an element, compound, or other material (if it occurs in nature)
- Identifying the ways in which human activities define, perturb, or dominate material cycles (and establishing the magnitudes, trends, and causes of resource flows within an anthropogenically dominated system)
- Determining the environmental and resource supply implications of these perturbations

Once this information is in hand, focused, practical implications can be addressed: to mitigate undesirable environmental consequences related to human activities, we must have an accurate understanding of those activities and of how they might be changed. One form of change is largely technological, and involves the redesign of products and processes such that the use of materials is optimized; the environmental implications of manufacture, delivery, and customer use are minimized; and the eventual recovery and reuse of resources are enhanced. A second form of change is behavioral, and involves economic producers and consumers and the forces that determine their adoption of technologies that alter the use of materials.

A useful perspective on the intellectual challenges presented by technological material cycles is provided by activities related to the biogeochemical cycles of Grand Challenge 1. For those cycles, natural and perturbed, the research activities are centered on identifying the complete suite of sources, sinks, and feedback loops; assessing how these variables have evolved over time; and predicting how they are likely to evolve in the future. Complicating factors include missing or poorly quantified information, incomplete understanding of human activities that shape the budgets, substantial spatial variation, and uncertainty about the behavior of the sources and sinks under altered physical and chemical conditions.

Many of these difficulties are present as well for the budgets controlled largely by anthropogenic activity. While sources are often rather well established, both in kind and magnitude, sinks and feedback loops are not, and the forms and magnitudes of storage in various reservoirs (formal and informal stockpiles, landfills, environmental receptor basins) are generally quite uncertain. In addition, the human activities that shape the budgets are not well documented or understood. In many cases, proxy data, inference, and archeomaterials research will be necessary to complete the picture. In this connection, the most ambitious portion of this grand challenge activity is likely to be the acquisition, comprehension, and integration of data sets and other information from the environmental, economic, and social spheres, and the development of robust ways of utilizing those results in predictive exercises. The achievement of data harmony, consistency, and rigor across this interdisciplinary landscape will be a major effort and will provide the necessary basis for a scientific understanding of material cycles.

An understanding of the use of materials and its implications is a prerequisite for many of the predictive exercises encompassed by the other grand challenges. As one example, Grand Challenge 5, Infectious Disease and the Environment, identifies chemical selection pressure from the environment on pathogens as a priority research area. This area could make use of data on contemporaneous rates of heavy metal and pesticide loss to ecosystems of interest, as well as informed projections of how those flows might be expected to change over space and time. A second example relates to Grand Challenge 4, Hydrologic Forecasting. Informed projections derived from analysis of material use would be directly applicable to predictions of water availability and quality.

Scientific Readiness

This grand challenge is timely for both scientific and policy reasons. From a scientific standpoint, work has begun on devising regional and global budgets for several of the toxic trace metals (e.g., Jolly 1992, Jasinski 1995). These and related studies have started identifying data sources related to extraction, processing, use, and disposal, and provide a framework for more general research related to the budgets of key materials used by humanity. Moreover, the sophisticated techniques and considerable scientific expertise developed to investigate nutrient cycles are directly applicable to questions about material cycles, and thus can be used to initiate research efforts in this area. In addition, as part of the Industrial Transformations project of the International Human Dimensions Programme (1999), social science and policy research has begun to address changes in production and consumption patterns. This effort is developing an international and interdisciplinary research community that is addressing fundamental questions about consumption trends and their causes that must be addressed to predict future trends in material cycles and the environmental effects of these changes.

Historical changes in material mobilization, use, and dissipation are beginning to be understood, for example, by constructing histories of fossil fuel consumption or trace metal deposition on polar ice or lake sediments. In addition to providing historical information on material utilization and dispersal, these data contribute to understanding of the historical intensity of interactions between human activities and the environment. Such efforts need to be expanded to include a wide range of materials and locales, with the ultimate goal of constructing gridded budgets integrated over the time period since the Industrial Revolution. International collaborative efforts should be encouraged, since material cycles do not respect national boundaries.

In a more practical vein, engineers in industry and academia are beginning to devote significant effort to "design for environment," in which the selection, processing, and use of materials play central roles. This technology-oriented research is key to the implementation of insights gained from an understanding of reservoir contents and flows, and could lead to reinvention of the ways in which materials are acquired and used by modern technology. One could envision, for example, that the results would stimulate the development of policy instruments designed to encourage the recovery, reprocessing, and reuse of a variety of selected materials, along with the development of technologies that would make these policy instruments implementable in efficient and effective ways. Such new approaches to material utilization would be informed by research in the environmental sciences in general and materials-environment interactions in particular, and enabled by modern engineering tools such as life-cycle assessment, computer-aided design and manufacturing, and performance analysis.

Important Areas for Research

1. Develop spatially explicit budgets for selected key materials. This research would involve quantifying reservoir contents and flows for the materials in question; constructing spatially resolved maps of these stocks and flows; and combining these results with other environmental, economic, and social data sets to learn more about the causes and consequences of changes in material cycles. As has been demonstrated by budgets for naturally cycling compounds of carbon and nitrogen, budgets constructed with a high degree of spatial resolution are much more useful than those that provide only aggregate, global information. The budgets thus developed would include analyses of anthropogenic flows by type of activity (e.g., mining, manufacturing, household use) and by technology, as well as by spatial location. They would require as well comprehensive integration with data on natural flows of the same materials. The generation of location-specific information would provide links between anthropogenic material cycles and their human causes and potential environmental impacts.

2. Develop methods for more complete cycling of technological materials. Addressing this topic would involve pursuing life-cycle design of products; lengthening the useful life of products by modular design; and advancing research on the utilization of residue streams, the recovery of discarded materials, and the transformation of patterns of consumption. The work might also involve the use of more easily recycled and reused materials, perhaps including benign new materials such as biological products and composites. This is largely an engineering activity, but one whose priorities are established by contemporary budgets and future budget scenarios.

3. Determine how best to utilize materials that have uniquely useful industrial applications but are potentially deleterious to the environment. This research would include describing the spectrum of uses for these materials, identifying points of loss of the materials to the environment and methods by which such loss might be reduced, developing substitutes for these materials, and investigating reengineering activities that could be used to cycle the materials more completely. As with other types of material use, but especially in this case, dematerialization (accomplishing a given design goal with a substantially smaller amount of material) could contribute substantially.

4. Develop an understanding of the patterns and driving forces of human consumption of resources. This research would involve studying material consumption patterns across time, in different countries, and at different levels of economic activity, with the aim of understanding how differences develop, why the patterns change, and what changes might be anticipated in the next several decades. The results would aid in understanding current patterns of material flows and provide a basis for anticipating societal drivers of those flows in the future.

5. Formulate models for possible global scenarios of future industrial development and associated environmental implications. This research would draw on contemporary material budgets, predictions of technological developments, studies of consumption patterns, and assessments of industry structure and environmental law and policy to predict how specific circumstances or policy options might strongly influence industry-environment interactions in the next several decades. Thus, this research constitutes the equivalent for impacts of resource and material use of scenario exercises such as those of the Intergovernmental Panel on Climate Change (1996).

3

Recommended Immediate Research Investments

As discussed in Chapter 1, the committee was asked to identify a small number of areas of environmental science (perhaps three or four) that are especially deserving as the focus of new research initiatives. Applying the criteria outlined in Chapter 1, the committee judged four areas to have the highest priority for immediate research investment. These four areas represent actions that the NSF, preferably in cooperation with other relevant federal agencies, can begin to undertake immediately. All meet the criteria of scientific novelty and excitement, likelihood of a large practical payoff, feasibility, timeliness, and magnitude. As with the grand challenges, the committee did not attempt to set priorities among the four areas; rather, they are presented in the same order as the corresponding grand challenges in Chapter 2:

- *Biological Diversity and Ecosystem Functioning:* an initiative to develop a comprehensive understanding of the factors that generate, maintain, and diminish biological diversity and their effects on ecosystem functioning.
- *Hydrologic Forecasting:* an initiative to develop the capability for regional hydrologic forecasting, specifically including the ecological consequences of changing water regimes.
- *Infectious Disease and the Environment:* an initiative to develop a comprehensive ecological and evolutionary understanding of infectious and environmental diseases.
- *Land-Use Dynamics:* an initiative to develop a systematic, spatially explicit understanding of the changes in land use and land cover that are critical to ecosystem functioning, ecosystem services, and human welfare.

The following sections briefly describe these four recommended research investments and some of the key issues for each, including data needs, coordination with other environmental science research, and other key issues that must be addressed in undertaking the research. The key issues raised in this chapter are specific to or especially critical for particular research areas. Some additional issues apply to all the grand challenges and therefore arise with all four recommended research investments; these cross-cutting issues are discussed in Chapter 4.

The committee is enthusiastic about its recommendations for immediate action, but we recognize that the relative importance of scientific challenges and recommended actions in light of the above criteria will evolve over time. As new challenges emerge or significant progress is made in addressing existing ones, the priorities for action will require reevaluation. The same is likely to be true if a non-U.S. perspective is taken, since groups from other areas of the world might view the priorities differently. Therefore, we recommend that an evaluation similar to the present one be repeated at approximately 5-year intervals, perhaps in collaboration with international organizations.

RECOMMENDED IMMEDIATE RESEARCH INVESTMENT 1: BIOLOGICAL DIVERSITY AND ECOSYSTEM FUNCTIONING

Recommendation: Develop a comprehensive understanding of the relationship between ecosystem structure and functioning and biological diversity. This initiative would include experiments, observations, and theory, and should have two interrelated foci: (a) developing the scientific knowledge needed to enable the design and management of habitats that can support both human uses and native biota; and (b) developing a detailed understanding of the effects of habitat alteration and loss on biological diversity, especially those species and ecosystems whose disappearance would likely do disproportionate harm to the ability of ecosystems to meet human needs or set in motion the extinction of many other species.

This initiative is compelling because (a) understanding the relationship between species diversity and ecosystem functioning poses a great intellectual challenge, and would lead to both scientific and practical breakthroughs; (b) humans use a large proportion of the nonglaciated land surface of the Earth, as well as its marine resources, creating the likelihood that biotic reserves—even combined with environmental restoration—will not by themselves be sufficient to prevent the extinction of many species; (c) increasing human demand for ecosystem goods and services threatens to outpace ecosystems' capacity to sustain those supplies and to maintain natural diversity; (d) given recent advances in the science of biological diversity, we are poised to make breakthroughs in understanding how diversity has been generated and maintained in nature, as well as how civilization can continue to sustain that diversity; and (e) current research in this area is inadequate, especially in the major foci recommended here.

This initiative is drawn from the areas for research described in Chapter 2, but is focused more narrowly to emphasize those areas not currently receiving enough attention. It includes research into a broad range of ecosystems and habitats, from those managed exclusively for protection of native biota, to those providing ecosystem goods and services, to those designed to provide for human needs and preferences while simultaneously supporting biological diversity. Experiments designed to elucidate the relationship of ecosystem functioning to species diversity would build on the few such experiments conducted to date. Work on ecosystems that are as close to natural as possible would reveal how those ecosystems function, helping us clarify the principles that should be applied in designing new kinds of habitats that can serve both human needs and those of other species. Similar experimental research must be done on human-dominated ecosystems because they currently constitute a large proportion of what is available to manage. Information gained from systematic research into the environmental requirements of native species could then be combined with our new understanding of ecosystem functioning to offer a wide variety of novel opportunities for understanding and managing anthropogenic landscapes.

The overall effort will require interdisciplinary research involving ecologists, ethologists, psychologists, engineers, economists, planners, landscape architects, and others. The work should interact with recommended immediate research investment 4 on land-use dynamics. Moreover, since a substantial proportion of human habitation occurs in or near freshwater ecosystems, the potential for this effort to be informed by and to inform the recommended immediate research investment on hydrologic forecasting is particularly promising. Strong potential synergies also exist with recommended immediate research investment 3 on infectious disease, which directly addresses interactions between humans and the ecosystems they influence.

Data Needs

The definition of data needs and the collection and synthesis of data will require close cooperation among physical, biological, and social scientists; engineers and planners; and the associated funding agencies.

Coordination with Other Environmental Science Research

Coordination of this research with other efforts in environmental science will help in understanding the controls on and means of protecting biological diversity. The following specific linkages are recommended:

• Coordinate this research with work on hydrologic models of runoff and subsurface water, which reflect the way living things, including people, interact with the landscape (Grand Challenge 4).

• Incorporate effects of human resource management institutions on eco-systems (Grand Challenge 6).

• Incorporate the effects of changing patterns of land use and land cover on the potential for (and limitations of) habitat redesign (Grand Challenge 7).

• Incorporate the effects of climate variability in assessments of ecosystem functioning and in the design of habitats to buffer for disturbances and extreme events (Grand Challenge 3).

• Develop partnerships with urban Long-Term Ecological Research sites.

RECOMMENDED IMMEDIATE RESEARCH INVESTMENT 2: HYDROLOGIC FORECASTING

Recommendation: Establish the capacity for detailed, comprehensive hydrologic forecasting, including the ecological consequences of changing water regimes, in each of the primary U.S. climatological and hydrologic regions. Important specific research areas include all those described under Grand Challenge 4.

This initiative is compelling because (a) hydrologic systems (physical and biological) are widely recognized as extremely vulnerable; (b) seismic tomography, remote sensing, and geographic information systems provide dramatic new tools for acquiring hydrologic information; and (c) theoretical models are evolving to be capable of using the new information.

Five distinct climatological and hydrologic regimes are generally recognized in the contiguous United States: semi-arid (western region), desert (southwestern region), midlatitude (central region), humid subtropical (southeastern region), and humid continental (northeastern region). Each environment has a unique combination of precipitation, evapotranspiration, topography, hydrologic response, and biotic community. For each of these regions, research is needed in the following areas (described in more detail in Chapter 2):

• Improve understanding of hydrologic and geomorphic responses to precipitation.

• Improve understanding of surface water generation and transport.

• Examine environmental stresses on aquatic ecosystems.

• Explain the relationships between landscape change and sediment fluxes.

• Improve understanding of subsurface transport.

• Map groundwater recharge and discharge vulnerability.

Data Needs

Data needs for this research include the following:

- Global, high-resolution remote sensing measurements from satellites and aircraft
- Subsurface data collected through geophysical tomography techniques
- Hydrologic and ecological field observations and experiments that involve both remote-sensing data and ground-based measurements
- Data on human activities that affect hydrologic systems (e.g., water use, contaminant releases, land transformations)

Coordination with Other Environmental Science Research

Hydrologic forecasting would be greatly enhanced by incorporating analyses of human activities that affect hydrologic systems over the time scales covered by geohydrologic, geomorphological, and ecological models, such as changes in land use, water demand, and industrial and agricultural activities and practices. Thus, the practical value of these models could be greatly enhanced by linkages with research efforts on the following:

- Biological diversity and ecosystem functioning (Grand Challenge 2)
- Land-use dynamics (Grand Challenge 7)
- Use and dispersal of materials (Grand Challenge 8)
- Resource management institutions (Grand Challenge 6)
- Predictive models of climatic systems (Grand Challenge 3)
- Biogeochemical cycles that affect the entry of contaminants and nutrients into aquatic systems (Grand Challenge 1)
- Transport and modification of disease vectors and hosts by freshwater systems (Grand Challenge 5)

Other Key Issues

It will be essential to forge and maintain links between the scientists who collect hydrologic information and develop forecasts and the users of those forecasts.

RECOMMENDED IMMEDIATE RESEARCH INVESTMENT 3: INFECTIOUS DISEASE AND THE ENVIRONMENT

Recommendation: Develop a comprehensive ecological and evolutionary understanding of infectious diseases affecting human, plant, and animal health.

This initiative is compelling as an immediate research investment because (a) it would promote linkages among scientific disciplines that are needed to understand critical environmental phenomena; (b) the current structure of support for environmental science is unlikely to cover the needed research; and (c) the broad availability of analytic tools makes the time ripe for significant progress in this area.

This initiative essentially involves developing a new interdisciplinary field. The focus of this field should be on the effects of physical, biological, chemical, climatic, and human processes as selective agents on pathogen virulence and host resistance; the impacts of environmental change on disease epidemiology and toxic organisms; methods of surveillance and monitoring; and the development of theoretical models of host-pathogen ecology, genetics, and evolution.

Data Needs

The data needs in this research area include the following:

- Enhanced surveillance of disease prevention and incidence in human and other target host populations
- Development of a molecular/genetic taxonomy of responsive organisms
- Sequencing of genomes of major pathogens

Coordination with Other Environmental Science Research

Coordination of this work with other efforts in environmental science should include the following specific linkages:

- Integrate this research initiative into existing programs by linking it to ecosystem-based research (e.g., studies of the Chesapeake Bay and some Long-Term Ecological Research sites).
- Expand large population-based prospective studies of disease (e.g., malaria and tuberculosis) to include information on environmental determinants of population health (such as airborne particles and gaseous copollutants, or pesticides and trace metals in food), thus achieving economies of scale and generating useful examples for the design of further research.
- Forge links to the U.S. Global Change Research Program, particularly research on climate change modeling.
- Integrate knowledge from research on other grand challenges: biological diversity and ecosystem functioning (Grand Challenge 2), land-use dynamics (Grand Challenge 7), climate variability and its effects on humans and natural systems (Grand Challenge 3), resource management institutions (Grand Challenge 6), and hydrologic forecasting (Grand Challenge 4).

Other Key Issues

There are two additional key issues to be addressed for this research initiative:

• To develop this research investment, it is critical to overcome the artificial separation of environmental scientists from biomedical and public health scientists, which is reinforced by the traditional structures of disciplinary affiliations and federal agency missions.

• Overcoming this separation will involve supporting interdisciplinary training of young scientists, providing funding incentives to encourage collaborative research among scientists in the relevant fields, and increasing the coordination/interaction among the federal science agencies that support research in health and environmental sciences.

RECOMMENDED IMMEDIATE RESEARCH INVESTMENT 4: LAND-USE DYNAMICS

Recommendation: Develop a spatially explicit understanding of changes in land uses and land covers and their consequences.

This initiative is compelling because (1) land use dominates many interactions between humans and the environment, and (2) the growth of knowledge in this area has been severely hampered by inadequate funding.

A successful approach to this initiative will require the development of long-term, regional databases for land uses, land covers, and related social information. It will also require (and encourage) innovative applications of dynamic spatial simulation.

Data Needs

The following issues are associated with the data needs for this research initiative:

• Costs to researchers of data from satellite imagery need to be reduced.

• The acquisition of cloud-free observations with good temporal and spatial coverage must be improved.

• Ground data on land cover need to be collected systematically and accurately.

• Data attributes and formats must be standardized across regional data centers.

• Standardized social, political, and economic data are required to test new theory and create the necessary new generations of models. Collection of such data at fine resolution will also facilitate comparative research and data sharing among research groups.

Coordination with Other Environmental Science Research

Understanding of patterns and trends in land-use change would benefit from coordination with research on the following:

- Climate variability and its effects on humans and ecosystems (Grand Challenge 3)
- Hydrologic forecasting (Grand Challenge 4)
- Biological diversity and ecosystems, especially research into human preferences for environmental conditions (Grand Challenge 2)
- Effects of institutions on natural resources (Grand Challenge 6)

Other Key Issues

Support is needed for critical international data collection and data harmonization.

MOVING FORWARD

Implementation of the above four recommended research investments should be facilitated by the recommendations of NSF's National Science Board (2000). The latter recommendations include an increase in funding for environmental research of $1 billion per year, phased in over the next 5 years; increased interdisciplinary research; increased support for long-term research; and implementation partnerships between NSF and other federal agencies. Those recommendations complement the ones offered in this report; thus implementation of the two sets of recommendations should be synergistic.

Planning Workshops

Although the committee has chosen four areas for immediate research investment, its membership did not encompass the expertise and lacked the resources needed to define the details of the individual research programs. Indeed, such definition is properly left to the federal agencies and scientific communities involved. To that end, the committee recommends that NSF, with other agencies as appropriate, convene workshops for each recommended immediate research investment to discuss and plan the research agenda, and that it consider convening similar workshops for each of the other important areas for research outlined under the grand challenges in Chapter 2. Such workshops, which might need to be repeated, each would involve 25-30 participants, many with broad interdisciplinary backgrounds and experience, and would include significant representation from both the natural and social sciences.

The workshops should include research scientists in academia, the relevant agencies, and the private sector, as well as potential users of the research results. Involvement of potential users, including public officials, private and nonprofit organizations, and interested and affected members of the public, would help inform the scientific community about user needs so that attention could be directed toward producing useful results where scientific capabilities make this possible. (This issue is discussed in more detail in Chapter 4.) The workshops should also include scientists whose primary interests lie in other grand challenge research areas that relate to the area under discussion (as indicated in the preceding sections). Their presence would facilitate research linkages across the grand challenges and help achieve economies in the overall research portfolio. For example, researchers might be able to suggest simple modifications in the data collection strategy for one area that would provide valuable information for those working in related areas. The workshops should produce reports to NSF, with other agencies as appropriate, that include the following components:

- A 5-year plan for the implementation of an initial detailed research program. Anticipated results in both basic research and information useful to society within a reasonable time period should be addressed.
- Consideration of the degree to which existing institutions in their present or modified form could play integral roles in the program.
- A program for training the necessary individuals in areas in which appropriate scientists are in short supply.
- A strategy for developing the necessary research integration across disciplines. Various approaches to this end should be considered, including training interdisciplinary scientists, encouraging environmental scientists to collaborate across disciplines, and strengthening interdisciplinary research communities.
- Discussion of the usefulness and importance of regional approaches and integrated laboratories for advancing the specific area of research and, if considered important, recommendations on the appropriate institutional form for such laboratories.
- A strategy for coordinating the program results with those of other grand challenge initiatives.
- A strategy for enhancing (and evaluating) the usefulness of the scientific information to be generated.
- An estimate of the financial and other support (e.g., data availability) needed to implement the 5-year plan.

Funding Requirements

Until the reports from the workshops outlined above are available, it will not be possible to provide detailed estimates of the funding needed to implement the recommended immediate research investments. However, the committee judged it important to provide an estimate of the order of magnitude of support likely to be required to achieve progress on the various research investments. To this end, we estimated for each initiative how many researchers (university faculty, graduate students, postdoctoral fellows) would be required to make substantial progress. We then considered major infrastructure needs (e.g., satellites, major laboratory facilities). The various initiatives have differing needs, and several of the major expenses apply to more than one initiative; therefore, the committee's estimates are only approximations. Given these caveats, the committee estimates that each of the recommended immediate research investments would require several hundred million dollars (at a minimum) over a 10-year period, for a total investment of perhaps $1-2 billion for the four initiatives combined. This investment of $100 or $200 million per year is well within the National Science Board's (2000) recommended budget increase for environmental sciences of $1 billion per year.

4

Implementation Issues

Progress on the next generation of environmental research will depend on dealing successfully with several important implementation issues. These issues should be addressed in a topic-specific manner in the planning workshops recommended in Chapter 3 and should be considered in overall planning of environmental research within NSF. The committee considers all of these issues to be important to environmental research, but does not recommend that they be addressed in a uniform way across all research fields. Indeed, in one case (the need for regional research centers), there was considerable diversity of opinion among committee members.

COMPLEXITY OF ENVIRONMENTAL PHENOMENA: AN OVERALL RESEARCH VISION

The grand challenges set forth in this report cannot be pursued effectively in isolation from each other because they are closely interrelated. For example, changes in ecosystems, biological diversity, hydrologic systems, and pathogen-host relations are all affected by climate change. Changes in ecosystems and hydrologic systems can also affect climate and are affected by changing patterns of land use and increasing use of materials by human populations. Biogeochemical cycles, hydrologic systems, and ecosystems are all affected by and can affect climate, use of materials, and the institutions that shape human use of natural resources. In short, most of the phenomena central to each grand challenge act as driving variables for phenomena at the center of other grand challenges. In addition, actions intended to affect one environmental system may simultaneously affect a variety of other systems. This phenomenon is familiar to environmental regulators as the problem of cross-media relationships (for example, regulations on water pollutants may lead to increases in toxic releases into the air or land). The phenomenon is generic to changes in environmental systems and presents major analytical difficulties.

Perhaps even more challenging for science is that the outcomes of interest within each grand challenge depend simultaneously on change in more than one driving variable. The grand challenges require problem-oriented science that can integrate physical, biological, chemical, and human systems well enough to predict the response of critical regions or phenomena to multiple causal variables, sometimes referred to as multiple stresses. Understanding the interactions of these systems is imperative, because the many environmental factors now undergoing change make it difficult to assess the impact of any single change in the Earth system (particularly changes in human activities), and thus it is difficult to assess the outcomes of specific mitigation and adaptation strategies.

Understanding how environmental and human outcomes are affected by multiple driving variables lies beyond the capacity of any single environmental science discipline. Studies focused on single causal variables are typically inadequate and potentially misleading. As emphasized throughout this report, the needed understanding will require true integration of the social sciences and engineering, as well as various disciplines within the natural sciences, around common research problems. Expertise can be drawn from many institutions across the country to focus on research within a specific region, but to work effectively, the research recommended here will have to involve new kinds of scientific teams and communities capable of communicating and collaborating across the natural science-social science gulf. These groups will require a large number of scientists with broad, interdisciplinary perspectives, as well as an increased capability for cross-disciplinary collaboration among environmental scientists, who may develop more interdisciplinary orientations as a result.

Science is becoming increasingly capable of developing the observational basis, focused process studies, and coupled models needed to provide a firm foundation for considering multiple causal factors (multiple stresses) in environmental analysis and assessment. A useful strategy for developing this capability in the context of meeting the grand challenges is to analyze environmental phenomena in "natural laboratories." These would typically be regions in which key environmental perturbations are occurring, and on which there exists a base of information that can be organized to provide the foundation for a model of the ecological, biophysical, and human systems to be studied. Natural laboratories can bring together researchers from different scientific fields around common research problems. They can also make possible concerted research on several of the grand challenges, taking advantage of the fact that some of the same observations and models will be useful for several lines of research.

A systems model can organize research and ensure communication and collaboration among participating scientists. Research in natural laboratories would aim to develop regional models capable of projecting the future of the system under study, including effects of change on ecological and social outcome variables and of human activities on environmental systems. Such models should allow for exploration of the likely outcomes under continuation of existing conditions or under change in forcing factors, and of the anticipated results of various adaptation strategies and attempts to mitigate change in one or more variables.

Examples of possible regional natural laboratories include the following:

• Major estuarine systems, such as the Chesapeake Bay, which are subject to a wide variety of stresses. These stresses include severe weather; climate variability; climate change; land-use changes that modify stream run-off patterns and sediment loading; human modification of river systems (e.g., dams); pollutants from agricultural, industrial, and urban regions; nutrient loading; resource use; sea-level change; invasive species; human modification of the adjacent shore; and disease. The combination of stresses renders problematic predictions of key environmental indices, such as oxygen levels or productivity, disease outbreaks (e.g., *Pfiesteria*), or changes in species composition.

• The developing megacities, especially along coastlines, which create a complex interplay between extensive human modification of the environment and human quality of life. The human impacts are numerous, including extensive land-cover change (often at the expense of soils and other ecosystem resources), extensive water needs in the surrounding regions, large-scale waste production, urban heat island effects, and modification of air quality. Severe storms, climate variability, and sea-level change add significant physical stresses to the system as well. The quality of life within major urban areas is also an issue, given the effects of climate change on mortality and morbidity from exposure to extreme heat and cold; links between respiratory illness and air quality; and a host of stresses related to poverty, crime, population pressures, aging infrastructure, and a variety of public health issues. Scientists currently lack the capability to examine megacities and their growth as integrated systems.

The key to future environmental research will be to develop a capability to examine such regions comprehensively, instead of examining one variable or issue at a time. The concept of integrated laboratories is one example of a mechanism for moving beyond individual disciplinary challenges to develop a strongly multidisciplinary research capability. The keys are (a) to develop a comprehensive set of physical, biological, chemical, and human system observations or "sensor webs" designed to gain understanding, aid model development, and validate predictions of coupled models; (b) to focus combined field and model process studies on areas or topics of critical uncertainty; and (c) to construct increasingly comprehensive regional system models in which the discipline of forecasting, evaluation, and improvement is rigorously applied. The development of comprehensive observations and models should be a major catalyst for multidisciplinary research, while common scientific objectives are likely to engender new modes and avenues of research. The emphasis on a region-specific predictive capability will drive the development of enhanced understanding and suites of high-resolution models that are likely to provide new capabilities to address a broad range of regional, national, and global environmental issues. The new urban Long-Term Ecological Research sites may be among the appropriate venues for integrated research on a regional basis involving the full range of environmental sciences.

IMPLEMENTATION OF REGIONAL APPROACHES AND THE ROLE OF RESEARCH CENTERS

The committee agreed that some of its research recommendations require or would benefit from a regional focus. For example, as noted in Chapter 3, the initiative on hydrologic forecasting needs to address five distinctive regional climatological and hydrologic regimes within the United States: semi-arid (western region), desert (southwestern region), midlatitude (central region), humid subtropical (southeastern region), and humid continental (northeastern region). Each of those regions has a unique combination of precipitation, evapotranspiration, topography, and hydrologic response. Similarly, the initiative on land-use and land-cover change will depend on developing regional databases, observatories, and archives, and the natural laboratories described above have a strong regional flavor.

The committee members did not agree, however, on how best to implement a regional focus, and particularly on whether to recommend the establishment of regional research centers. Some members argued that learning in regional natural laboratories cannot be adequately achieved without the interdisciplinary social and professional environment provided by a shared physical location. Those members argued that regional centers would act as nodes for intellectual organization and innovation and would be ideal sites for providing interdisciplinary training and for increasing the capability of environmental scientists to collaborate effectively on cross-disciplinary problems. They also argued that the visibility of centers as concrete entities would help attract funds for new research initiatives.

Other committee members did not favor recommending the establishment of centers. They argued that the National Research Council should not tell funding agencies in such detail how to accomplish the recommended research tasks. They also argued that large investments in centers could reduce the overall quality of research by allocating to facilities funds that might better be used for research, and by making it difficult for new ideas from researchers outside the centers to receive support. In-house competition for a center's funds could result in less rigorous proposal review and therefore in lower-quality research as compared with national research competitions. These committee members concluded that coordination can be achieved by specifying the needed cross-disciplinary collaborations in program announcements and by providing relatively low-cost support for meetings and conferences.

The committee concluded that the decision on whether to support bricks-and-mortar regional research centers should be made by the funding agencies on the basis of the scientific, capacity-building, and infrastructure needs associated with studying specific environmental systems. In making decisions about the institutional form for regional research in individual research areas, funding agencies should make use of the workshops described at the end of Chapter 3. We recommend that each of these workshops consider the usefulness and importance of regional approaches and integrated laboratories for advancing the specific area of research, and if such approaches are considered important, that the appropriate institutional form for such laboratories be considered as well.

The interrelationships among the grand challenges make it necessary for NSF to consider ways of supporting integrated research efforts that can help develop the observations, process studies, and models needed to investigate problems of multiple causal variables, cross-media relationships, and linkages across the grand challenge areas. Research centers or virtual laboratories focused on single problems such as hydrologic forecasting or biological diversity may not range broadly enough to build an adequate capability for such multiple-system investigations. NSF should therefore consider supplementary support mechanisms. One possibility would be to hold a competition for research centers or teams focused on particular multiple-variable problems outside or cutting across areas in which a major research investment is being made. Another would be to define coterminous regions for centers or virtual laboratories working in different grand challenge areas, and to support shared data and model development among them. A third would be to define sets of environmental, social, and economic indicators needed for studying multiple-variable issues in the environmental sciences, and to invest in the observational systems needed to close the distance between existing and needed data collection.

BUILDING OF CAPACITY FOR INTERDISCIPLINARY, PROBLEM-ORIENTED RESEARCH

Because of the nature of the phenomena at the center of the grand challenges, efforts to meet each challenge will benefit from interdisciplinary analysis. Whereas multidisciplinary research is a collaboration among investigators from different scientific fields, interdisciplinary research entails the integration of multidisciplinary knowledge. Nonadditive relationships and mutual causation among the variables studied in different disciplines make integration across disciplines highly desirable (e.g., Wijkman 1999, Clark 1999). However, interdisciplinary research and training have their costs as well as their potential benefits for environmental problem solving (e.g., Hansson 1999, Lasswell 1970). Balancing of the costs and benefits of interdisciplinary research and training was beyond the scope of the committee's work. But because the topic is relevant and important, we describe some current obstacles to producing true interdisciplinary research and some possible methods for overcoming them.

Integrated, interdisciplinary environmental research will require scientists with broad, interdisciplinary perspectives, as well as an increased capability among environmental scientists in a given discipline to understand enough about other disciplines to work fruitfully with scientists in those fields. Such research may also require strengthening of interdisciplinary research communities (e.g., through interdisciplinary professional meetings, associations, journals, summer training institutes), particularly in the environmental social sciences.

There are relatively few broadly interdisciplinary environmental scientists available to tackle the grand challenges outlined in this report. To utilize the talents of those interdisciplinary natural and social scientists, to increase their numbers, to encourage environmental scientists to collaborate across disciplines on cross-disciplinary problems, and to build interdisciplinary research communities, it will be necessary for funding structures to free individuals from the constraints imposed by disciplinary departments within universities and by the disciplinary panels that judge research proposals within funding agencies. It might be necessary to go beyond removing constraints. Mechanisms to be considered include forming interdisciplinary review panels; establishing mechanisms that will foster ongoing interdisciplinary collaboration (e.g., centers, laboratories, coordinators, virtual associations); funding networks for communication across research groups; supporting interdisciplinary communities; and even mandating integrated research across a range of relevant disciplines in calls for proposals, as has been done, for example, in the research program on water and watersheds of NSF, the Environmental Protection Agency, and the Department of Agriculture. It may be advisable to adopt multiple mechanisms because of the variety of barriers to interdisciplinary research.

Training is particularly important, especially for producing a new generation of interdisciplinary scientists, but also for improving the capabilities of scientists to work effectively in multidisciplinary teams. Universities are generally organized according to traditional disciplines, posing barriers to interdisciplinary research and training. While innovative departments and institutes have been established at some universities, they are few in the United States. It is still unusual to find a program that trains students in several of the relevant natural science and social science fields.

One mechanism that can provide a cross-disciplinary learning environment for undergraduate and graduate students is support for interdisciplinary research training groups. A training grant centered on a grand challenge could bring interested students to a university for periods of a few months to several years, and could sponsor such activities as visiting speakers, symposia, and workshops that would bring together faculty and students from several different departments. In addition, a training grant could provide funds for equipment and facilities related to a research challenge. Training grants might also support summer programs that would attract graduate students and faculty to a single location for courses on new research techniques or data sets. In general, training grants are inexpensive compared with centers or institutes, and they have a built-in sunset clause since their existence depends on funding, rather than on a structural change in the university. The committee recommends that each of the planning workshops described in Chapter 3 address the issue of how best to build the needed capacity for research integration across disciplines in its particular area of research.

NEED FOR INTERAGENCY SUPPORT OF GRAND CHALLENGE RESEARCH

Funding for multidisciplinary and problem-oriented research presents two important implementation issues within federal agencies. One is the tendency in some agencies to fund and review research by discipline, essentially replicating the traditional structure of universities. Thus, a proposal for such a research effort would not fare well if judged only by disciplinary review panels. An example may be the grand challenge involving an ecological and evolutionary understanding of infectious diseases, because the topic crosses the boundaries between disciplines and between the traditional purviews of NSF and the National Institutes of Health. If considered by only one of those agencies, the research might fail to achieve some of its promise to bring ecology and biomedical science together.

A second issue that might arise is due to the split between the perceived functions of so-called "research" and "mission" agencies. A sharp division between these designations is unfortunate because basic research in the environmental sciences is often inspired by practical needs. Agencies with resource management responsibilities need support from the environmental sciences to do their jobs well. On the other hand, research agencies that use public funds to support environmental research understand that the research should have some ultimate relevance for addressing environmental problems. Collaborations among both kinds of agencies on the grand challenges, such as apparently successful collaborations between NSF and mission agencies in funding environmental research on watersheds, industrial transformations, and other issues, could add depth and insight to the research and its results. The collaborating agencies will need to find ways to foster interdisciplinary collaboration and design research programs that adequately meet both curiosity-driven and decision-driven research needs.

NEED TO IMPROVE THE USEFULNESS OF ENVIRONMENTAL SCIENCE RESEARCH

Investments in the grand challenges will yield both scientific and practical payoffs as outlined in Chapter 2. However, major environmental science efforts of the past have not always had the strong practical value promised by proponents. For example, the National Acid Precipitation Assessment Program was prominently criticized as a good science project that yielded little information of use for policy (Rubin et al. 1992). Risk assessments for nuclear power plant operations, radioactive waste disposal, dioxin exposure, and other hazards have cost billions of dollars over many years, but have not resolved the scientific issues of greatest concern to participants in policy decisions (National Research Council 1994, 1996). The U.S. Global Change Research Program may have learned from such experiences. It invited regional participants in the 1998-1999 national assessment of climate change to discuss the relevance of scientific information resulting from the program to the needs of local decision makers, and the program has taken new directions as a result. It is important for research on the grand challenges to do well at responding to the informational needs of practical decision makers. However, doing so will itself require coordinated research focused on identifying and addressing the needs of decision makers and helping scientists make their contributions more understandable and relevant to the decisions being made.

Research on human response to environmental science information reveals some of the reasons for past failures and offers lessons for future research programs (National Research Council 1989, 1996, 1999c). One reason for failure is that new scientific information may not fit well into people's usual modes of understanding and may therefore be ignored or systematically misinterpreted (Fischhoff 1994, 1998; Fischhoff and Downs 1997; Kahneman et al. 1982; Slovic 1987). Overcoming this problem requires systematic efforts to understand how people think about the relevant environmental processes and to develop information accordingly. The problem of achieving understanding is likely to be especially serious when the scientific information comes from complex system models yielding counterintuitive results.

A second reason for failure results from the reliance of most users of scientific information on intermediaries, not scientists, for interpretations of research results. These intermediaries include mass media organizations, political commentators and interest groups, trade associations, social movement organizations, insurers, law firms, consultants, and government bureaucracies at all levels. When environmental scientists write reports and make public statements, they typically do not consider whether effective intermediaries are in place to reach the intended audiences, or whether existing intermediaries may ignore, shade, misinterpret, or deliberately distort the scientific conclusions. Although the design of effective information-delivery systems usually lies outside the expertise and interest of environmental scientists, it is important for making environmental science information useful. In particular, the design must be sensitive to the needs and capabilities of its intended audience (Jones et al. 1998, 1999).

A third reason that environmental science may not live up to its practical potential is that the research questions addressed by scientists may not be those for which decision makers most want answers. For example, climate modelers may do excellent research to predict average precipitation, while planners want information on the likelihood of extreme precipitation events (e.g., Policansky 1977); risk assessors may predict the incidence of cancer in an entire population, while public health officials may be most concerned with risks to children. Sometimes the science does not match informational needs because theory and knowledge are insufficient to yield the desired information. Sometimes, however, having a clear picture of the needs of decision makers, including public officials, private and nonprofit organizations, and interested and affected members of the informed and attentive public can allow the scientific community to develop more relevant information than would otherwise be the case. Dialogue between environmental scientists and those whose lives the science is intended to improve can help uncover such possibilities for mutual benefit and clarify the limitations of science for those who want information that lies beyond present scientific capabilities. In so doing, dialogue may also help ensure the trusting relationship needed if public support for environmental science is to grow and if the information science produces is to be deemed credible. Such dialogue is typically required from the beginning of a research program, when the scientific questions are being framed (Fischhoff 2000, Institute of Medicine 1999, National Research Council 1996). It is for this reason that the users of environmental research should be included in the planning workshops recommended in Chapter 3. Some federal agencies have been experimenting with such dialogues and report that the usefulness of the science improves without its quality being compromised.

Increasing the usefulness of research may also require research to identify the kinds of information that could benefit various types of decision makers, the information they want, and the modes of presentation and systems of information delivery that would facilitate their effective use of the information. It may also require research, sometimes called "translational," that establishes the implications of knowledge about basic processes for practical applications.

References

Abdel-Wahab, M. F., G. T. Strickland, A. El-Sahly, N. El-Kady, S. Zakaria, and L. Ahmed. 1979. Changing pattern of schistosomiasis in Egypt 1935-79. Lancet 2(8136)242-244.

Agrawal, A. 1999. Greener Pastures: Politics, Markets, and Community among a Migrant Pastoral People. Duke University Press, Durham, NC.

Allenby, B. R. 1999. Industrial Ecology: Policy Framework and Implementation. Prentice Hall, Upper Saddle River, NJ.

Alves, D. S., and D. L. Skole. 1996. Characterizing land cover dynamics using multi-temporal images. International Journal of Remote Sensing 17(4):835-839.

American Meteorological Society. 1998. Special Issue on NCAR's Climate System Model. Journal of Climate 11(6).

Amthor, J. S. 1995. Terrestrial higher-plant response to increasing atmospheric [CO_2] in relation to the global carbon cycle. Global Change Biology 1(4):243-274.

Arrhenius, S. 1896. On the influence of carbonic acid in the air upon the temperature of the ground. Philos. Mag. 41:237-275.

Bailey, T. C., and A. C. Gatrell. 1995. Interactive spatial data analysis. Longman Scientific and Technical, Harlow, UK.

Bassett, T., and Z. Koli Bi. 2000. Environmental discourses and the Ivorian Savanna. Annals of the Association of American Geographers 90(1):67-95.

Baumol, W. J., and W. E. Oates. 1988. The Theory of Environmental Policy, 2nd ed. University Press, Cambridge, U.K.

Benkovitz, C., M. Scholtz, M. Trevor, and T.E. Graedel. 1996. Global gridded inventories of anthropogenic emission of sulfur and nitrogen. Journal of Geophysical Research 101:29239-29253.

Berkes, F., and Folke, C. 1998. Linking Social and Ecological Systems: Management Practices and Social Mechanisms for Building Resilience. Cambridge University Press, New York.

Blomquist, W. 1992. Dividing the Waters: Governing Groundwater in Southern California. ICS Press, San Francisco.

Bockstael, N. E. 1996. Modeling economics and ecology: The importance of a spatial perspective. American Journal of Agricultural Economics 78:1168-1180.

Bolin, B., and R. B. Cook. 1983. The Major Biogeochemical Cycles and Their Interactions. SCOPE 21. JohnWiley and Sons, Chichester, U.K.

Brasseur, G. P., D. A. Hauglustaine, and S. Walters. 1996. Chemical compounds in the remote Pacific troposphere: Comparison between MLOPEX measurements and chemical transport model calculations. Journal of Geophysical Research 101:14795-14813.

Braswell, B. H., D. S. Schimel, E. Linder, and B. Moore III. 1997. The response of global terrestrial ecosystems to interannual temperature variability. Science 278(5539):870.

Bredehoeft, J. D. 1984. Chestel Kisiel Lecture, Water Management in the United States—A Democratic Process. University of Arizona, Tucson, AZ.

Bromley, D., ed. 1992. Making the Commons Work: Theory, Practice, and Policy. Institute for Contemporary Studies, San Francisco, CA.

Brown, J. S., and N. B. Pavlovic. 1992. Evolution in heterogeneous environments: Effects of migration on habitat specialization. Evolutionary Ecology 6:360-381.

Burkhart, J. G., G. Ankley, H. Bell, H. Carpenter, D. Fort, D. Gardiner, H. Gardner, R. Hale, J. C. Helgen, P. Jepson, D. Johnson, M. Lannoo, D. Lee, J. Lary, R. Levey, J. Magner, C. Meteyer, M. D. Shelby, and G. Lucier. 2000. Strategies for assessing the implications of malformed frogs for environmental health. Environ. Health Perspect. 108(1):83-90.

Burnham, K. P., and W. S. Overton. 1979. Robust estimation of population size when capture probabilities vary among animals. Ecology 60:927-936.

Chao, A., and S. M. Lee. 1992. Estimating the number of classes via sample coverage. J. Amer. Statistical Assoc. 87:210-217.

Chapin III, F. S., B. H. Walker, R. J. Hobbs, D. U. Hooper, J. H. Lawton, O. E. Sala, and D. Tilman. 1997. Biotic control over the functioning of ecosystems. Science 277:500-503.

Chazdon, R. L., R. K. Colwell, J. S. Denslow, and M. R. Guariguata. 1998. Statistical methods for estimating species richness of woody regeneration in primary and secondary rain forests of NE Costa Rica. In: F. Dallmeier, J. A. Comiskey, eds. Forest Biodiversity Research, Monitoring and Modeling: Conceptual Background and Old World Case Studies. Parthenon, Paris.

Chua, K. B., K. J. Goh, K. T. Wong, A. Kamarulzaman, P. S. Tan, T. G. Ksiazek, S. R. Zaki, G. Paul, S. K. Lam, and C.T. Tan. 1999. Fatal encephalitis due to Nipah virus among pig-farmers in Malaysia. Lancet 354(9186):1257-1259.

Clark, T. W. 1999. Interdisciplinary problem-solving: Next steps in the Greater Yellowstone Ecosystem. Policy Sciences 32:393-414.

Climate Monitoring and Diagnostics Laboratory. 1996-1997. Summary Report No. 24; NOAA Environmental Research Laboratory.

Cody, M. L. 1975. Towards a theory of continental diversities: Bird distribution over Mediterranean habitat gradients. Pp. 214-257 in M. L. Cody, and J. M. Diamond, eds. Ecology and Evolution of Communities. Harvard University, Cambridge, MA.

Colwell, R. R. 1996. Global climate and infectious disease: The cholera paradigm. Science 274(5295):2025-2031.

Crocker, T. D. 1966. The Structure of Atmospheric Pollution Control Systems. Pp. 61-86 in H. Wolozin, ed. The Economics of Air Pollution. W. W. Norton, New York.

Daily, G. C., (ed) 1997. Nature's Services. Island Press, Washington, D.C.

Daily, G. C. 1999. Developing a scientific basis for managing Earth's life support systems. Conservation Ecology 3:14 Web site http://www.consecol.org/vol3/iss2/art14/

Dales, J. H. 1968. Pollution, Property, and Prices. University of Toronto Press, Toronto.

Denton, J. S., S. P. Hitchings, T. J. C. Beebee, and A. Gent. 1997. A recovery program for the natterjack toad (Bufo calamita) in Britain. Conservation Biology 11:1329-1338.

Doggett, S. L., R. C. Russell, J. Clancy, J. Haniotis, and M. J. Cloonan. 1999. Barmah Forest virus epidemic on the south coast of New South Wales, Australia, 1994-1995: Viruses, vectors, human cases, and environmental factors. J. Med. Entomol. 36(6)861-868.

Driscoll, C. T., G. E. Likens, and M. R. Church. 1998. Recovery of Surface Waters in the Northeastern U.S. from Decreases in Atmospheric Deposition of Sulfur. Water, Air, and Soil Pollution 105(1/2):319.

DuPont, H. L., and J. H. Steele. 1987. The human health implication of the use of antimicrobial agents in animal feeds. Vet. Q. 9(4):309-320.

Environmental Protection Agency. 1998. Contaminated Sediment: EPA's Report to Congress. EPA-823-F-98-001. Washington, D.C.

Erickson III, D. J. 1999. Nitrogen deposition, terrestrial carbon uptake and changes in the seasonal cycle of atmospheric CO_2. Geophys. Res. Lett. 26:3313-3316.

Ewald, P. W. 1996. Guarding against the most dangerous emerging pathogens. Emerg. Infect. Dis. 2(4):245-257.

Fan, S., M. Gloor, J. Mahlman, S. Pacala, J. Sarmiento, T. Takahashi, and P. Tans. 1998. A large terrestrial carbon sink in North America implied by atmospheric and oceanic carbon dioxide data and models. Science 282(5388):442-446.

Fischhoff, B. 1994. What forecasts (seem to) mean. International Journal of Forecasting 10:387-403.

Fischhoff, B. 1998. Communicate unto others. Reliability Engineering and System Safety 59:63-72.

Fischhoff, B. 2000. Scientific management of science? Policy Sciences 33:73-87.

Fischhoff, B., and J. Downs. 1997. Overt and covert communication about emerging foodborne pathogens. Emerging Infectious Diseases 3:489-495.

Fleming, L. E., J. Easom, D. Baden, A. Rowan, and B. Levin. 1999. Emerging harmful algal blooms and human health: *Pfiesteria* and related organisms. Toxicol. Pathol. 27(5):573-581.

Franco, A. A., A. D. Fix, A. Prada, E. Paredes, J. C. Palomino, A. C. Wright, J. A. Johnson, R. McCarter, H. Guerra, and J. G. Morris Jr. 1997. Cholera in Lima, Peru, correlates with prior isolation of *Vibrio cholerae* from the environment. Am. J. Epidemiol. 146(12):1067-1075.

Frederick, K. D., and N. J. Rosenberg, sp. eds. 1994. Assessing the impacts of climate change on natural resource systems. Climatic Change 28(1/2), special editions.

Galloway, J. N. 1995. Acid deposition: Perspectives in time and space. Water, Air, and Soil Pollution 85:15-24.

Gibson, C. 1999. Politicians and Poachers: The Political Economy of Wildlife Policy in Africa. Cambridge University Press, New York.

Gibson, C., M. McKeon, and E. Ostrom. 2000. People and Forests: Communities, Institutions, and Governance. MIT Press, Cambridge, MA.

Glazovsky, N. F. 1995. The Aral Sea Basin. Pp. 92-139 in J. X. Kasperson, R. E. Kasperson, and B. L. Turner II, eds. Regions at Risk: Comparisons of Threatened Environments. United Nations University Press, New York.

Goovaerts, P. 1997. Geostatistics for natural resources evaluation. Oxford University Press, New York.

Graedel, T. E., and B. R. Allenby. 1995. Industrial Ecology. Prentice Hall, Englewood Cliffs, NJ.

Graf, W. L. 1993. Landscapes, commodities, and ecosystems: The relationship between policy and science for American rivers. Pp. 11-42 in Sustaining Our Water Resources, National Academy Press, Washington, D.C.

Graf, W. L. 1999. Dam nation: A geographic census of American dams and their large-scale hydrologic impacts. Water Resources Research 35(4):1305-1311.

Guenther, A., P. Zimmerman, and M. Wildermuth. 1994. Natural volatile organic compound emission rate estimates for U.S. woodland landscapes. Atmos. Environ. 28:1197-1210.

Hahn, B. H., G. M. Shaw, K. M. De Cock, and M. P. Sharp. 2000. AIDS as a zoonosis: Scientific and public health implications. Science 287:607-614.

Hansson, B. 1999. Interdisciplinarity: For what purpose? Policy Series 32:339-343.

Harvell, C. D., K. Kim, J. M. Burkholder, R. R. Colwell, P. R. Epstein, D. J. Grimes, E. E. Hofmann, E. K. Lipp, A. D. Osterhaus, R. M. Overstreet, J. W. Porter, G. W. Smith, and G. R. Vasta. 1999. Emerging marine diseases—climate links and anthropogenic factors. Science 285(5433): 1505-1510.

Hay, S. I., R. W. Snow, and D. J. Rogers. 1998. Predicting malaria seasons in Kenya using multitemporal meteorological satellite sensor data. Trans. R. Soc. Trop. Med. Hyg. 92(1):12-20.

Hess, C. 1999. A Comprehensive Bibliography of Common Pool Resources. CD-ROM. Indiana University, Bloomington. Workshop in Political Theory and Policy Analysis.

Holland, E. A., B. H. Braswell, J. A. Lamarque, A. Townsend, J. Sulzman, J. F. Muller, F. Dentener, G. Brasseur, H. I. Levy II, J. E. Penner, and G. Roelofs. 1997. Variations in the predicted spatial distribution of atmospheric nitrogen deposition and their impact on carbon uptake by terrestrial ecosystems. Journal of Geophysical Research 102:15849-15866.

Holt, R. D. and R. Gomulkewiecz. 1997. The evolution of species niches: A population dynamic perspective. Pp. 25-50 in H. G. Othmer, F. R. Adler, M. A. Lewis, and J. C. Dallon, eds. Case Studies in Mathematical Modeling: Ecology, Physiology, and Cell Biology. Prentice Hall, Upper Saddle River, NJ.

Houghton, R. A. 1994. The World Wide Effect of Land-Use Change. BioScience 44:305-313.

Houghton, R. A., J. L. Hackler, and K. T. Lawrence. 1999. The U.S. Carbon Budget: Contributions from Land-Use Change. Science. 285:574-578.

Hutchinson, G. E. 1959. Homage to Santa Rosalia, or why are there so many kinds of animals? American Naturalist 93:145-149.

Institute of Medicine. 1999. Toward Environmental Justice. National Academy Press, Washington, D.C.

Intergovernmental Panel on Climate Change. 1996. Climate Change 1995: The Science of Climate Change. J. T. Houghton, F. G. Meira Filho, B. A. Callander, N. Harris, A. Kattenberg, and K. Maskell, eds. Cambridge University Press, Cambridge, U.K. 570 pp.

International Geosphere-Biosphere Programme. 1997. The Terrestrial Biosphere and Global Change: Implications for Natural and Managed Ecosystems. B. Walker and W. Steffen, eds. IGBP Science No. 1, Stockholm: International Geosphere-Biosphere Programme.

International Geosphere-Biosphere Programme. 1999. The role of biospheric feedbacks in the hydrological cycle. Global Change Newletter (The IGBP-BAHC Special Issue) No. 39. Stockholm: International Geosphere-Biosphere Programme.

International Human Dimensions Programme on Global Environmental Change. 1999. IHDP Industrial Transformations Draft Science Plan, Free University of Amsterdam.

Jasinski, S. M. 1995. The materials flow of mercury in the United States. Resources, Conservation, and Recycling 15:145-179.

Jolly, J. H. 1992. Materials Flow of Zinc in the United States, 1850-1990. Open File Report 72-92, U.S. Bureau of Mines.

Jones, S., B. Fischhoff, and D. Lach. 1998. An integrated impact assessment for the effects of climate change on the Pacific Northwest salmon fishery. Impact Assessment and Project Appraisal 16:227-237.

Jones, S., B. Fischhoff, and D. Lach. 1999. Evaluating the usefulness of climate-change research for policy decisions. Climate Change 43:581-599.

Kahneman, D., P. Slovic, and A. Tversky, eds. 1982. Judgment under Uncertainty: Heuristics and Biases. Cambridge University Press, New York.

Karlquist, A. 1999. The meanings of interdisciplinarity. Policy Sciences 32:379-383.

Kasperson, J. X., R. E. Kasperson, and B. L. Turner II, eds. 1995. Regions at Risk: Comparisons of Threatened Environments. United Nations University Press, New York.

Keohane, R. O., and E. Ostrom, eds. 1995. Local Commons and Global Interdependence: Heterogeneity and Cooperation in Two Domains. Sage, London.

Kesler, S. E. 1994. Mineral Resources, Economics, and the Environment. Macmillan, New York.

Kiesecker, J. M., D. K. Skelly, K. H. Beard, and E. Preisser. 1999. Behavioral reduction of infection risk. Proceedings of National Academy of Sciences 96(16):9165-9168.

King, G. 1997. A solution to the ecological inference problem. Princeton University Press, Princeton, NJ.

Lambin, E. 1994. Modeling Deforestation Process: A Review. Trees Series B/11. Luxembourg: European Commission DGXIII.

Lanciotti, R. S., J. T. Roehrig, V. Deubel, J. Smith, M. Parker, K. Steele, B. Crise, K. E. Volpe, M. B. Crabtree, J. H. Scherret, R. A. Hall, J. S. MacKenzie, C. B. Cropp, B. Panigrahy, E. Ostlund, B. Schmitt, M. Malkinson, C. Banet, J. Weissman, N. Komar, H. M. Savage, W. Stone, T. McNamara, and D. J. Gubler. 1999. Origin of the West Nile virus responsible for an outbreak of encephalitis in the northeastern United States. Science 286(5448):2333-2337.

Lasswell, H. D. 1970. From fragmentation to configuration. Policy Sciences 2:439-446.

Lindsay, S. W., and M. H. Birley. 1996. Climate change and malaria transmission. Ann. Trop. Med. Parasitol. 90(6):573-588.

Linthicum, K. J., A. Anyamba, C. J. Tucker, P. W. Kelley, M. F. Myers, and C. J. Peters. 1999. Climate and satellite indicators to forecast Rift Valley fever epidemics in Kenya. Science 285(5426):397-400.

Liverman, D., E. F. Moran, R. R. Rindfuss, and P. C. Stern, eds. 1998. People and Pixels: Linking Remote Sensing and Social Science. National Academy Press, Washington, D.C.

Lobitz, B., L. Beck, A. Huq, B. Wood, G. Fuchs, A. S. Faruque, and R. Colwell. 2000. From the Cover: Climate and infectious disease: Use of remote sensing for detection of *Vibrio choerae* by indirect measurement. Proceedings of the National Academy of Sciences 97(4):1438-1443.

Loehman, E., and D. M. Kilgour, eds. 1998. Designing Institutions for Environment and Resource Management. Edward Elgar, Aldershot, U.K.

L'vovich, M. I., and G. F. White. 1990. Use and transformation of water systems. Pp. 235-252, in B. L. Turner II, W. C. Clark, R. W. Kates, J. F. Richards, J. T. Mathews, and W.B. Meyer, eds. The Earth as Transformed by Human Action: Global and Regional Changes in the Biosphere over the Last 300 Years. Cambridge University Press, Cambridge, U.K.

MacArthur, R. H., and E. O. Wilson. 1967. The Theory of Island Biogeography. Princeton University Press, Princeton, NJ.

Martens, W. J., L. W. Niessen, J. Rotmans, T. H. Jetten, and A. J. McMichael. 1995. Potential impact of global climate change on malaria risk. Environ. Health Perspect. 103(5):458-464.

Matson, P. A., R. L. Naylor, and I. Ortiz-Monastario. 1998. Integration of Environmental, Agrinomic, and Economic Aspects of Fertilization Management. Science 280:112-115.

McWhite, R. W., D. R. Green, C. J. Petrick, S. M. Seiber, and J. L. Hardesty. 1993. Natural Resources Management Plan, Eglin Air Force Base, Florida. U.S. Department of the Air Force, Eglin Air Force Base, FL.

Meyer, W. B., and B. L. Turner II, eds. 1994. Changes in Land Use and Land Cover: A Global Perspective. Cambridge University Press, Cambridge, U.K.

Meyer, W. B., and B. L. Turner II. 1992. Human population growth and global land-use/cover change. Annual Review of Ecology and Systematics 23:39-61.

Morison, W. L. 1989. Effects of ultraviolet radiation on the immune system in humans. Photochemistry and Photobiology 50:515-524.

Morris, J. G., Jr., and M. Potter. 1997. Emergence of new pathogens as a function of changes in host susceptibility. Emerg. Infect. Dis. 3(4):435-441.

Morse, S. S. 1993. Hantaviruses and the hantavirus outbreak in the United States. A case study in disease emergence. Ann. NY Acad. Sci. 740:199-207.

Moxness, E. 1998. Fisheries management: Not entirely a commons problem. Environmental Management.

Nabuurs, G.J., R. Paivinen, R. Sikkema, and G.M.J. Mohren. 1997. The role of European forests in the global carbon cycle—A review. Biomass Bioenergy 13(6):345-358.

Naeem, S. 2000. Reply to Wardle et al. Bulletin of the ESA 81(3):241-246.

Naiman, R. J., J. J. Magnuson, J. A. Stanford, and D. McKnight. 1995. The Freshwater Imperative. Island Press, Washington D.C.

Naiman, R. J. and M. G. Turner. 2000. A future perspective on North America's freshwater ecosystems. Ecological Applications 110(4):958-970.

National Aeronautics and Space Administration. 1999a. 1999 EOS Reference Handbook: A Guide to NASA's Earth Science Enterprise and the Earth Observing System. M. D. King and R. Greenstone, eds. Web site http://ltpwww.gsfc.nasa.gov/eospso/web/htdocs/eos_homepage/misc_html/refbook.html.

National Aeronautics and Space Administration. 1999b. EOS Science Plan. M. D. King, ed. Web site http://eospso.gsfc.nasa.gov/sci_plan/chapters.html.

National Drought Mitigation Center. 1999. Web site http://enso.unl.edu/ndmc/index.html/ University of Nebraska.

National Research Council. 1989. Improving Risk Communication. Committee on Risk Perception and Communication. National Academy Press, Washington, D.C.

National Research Council. 1992. Restoration of Aquatic Ecosystems: Science, Technology, and Public Policy. National Academy Press, Washington, D.C.

National Research Council. 1994. Science and Judgment in Risk Assessment. National Academy Press, Washington, D.C.

National Research Council. 1995. Science and the Endangered Species Act. National Academy Press, Washington D.C.

National Research Council. 1996. Understanding Risk: Informing Decisions in a Democratic Society. Committee on Risk Characterization. P.C. Stern and H.V. Fineberg, eds. National Academy Press, Washington, D.C.

National Research Council. 1997. Environmentally Significant Consumption: Research Directions. National Academy Press, Washington, D.C.

National Research Council. 1998. People and Pixels: Linking Remote Sensing and Social Science. Committee on the Human Dimensions of Global Change. D. Liverman, E. Moran, R. Rindfuss, and P. Stern, eds. National Academy Press, Washington, D.C.

National Research Council. 1999a. Adequacy of Climate Observing Systems. National Academy Press, Washington, D.C.

National Research Council. 1999b. Capacity of U.S. Climate Modeling to Support Climate Change Assessment Activities. National Academy Press, Washington, D.C.

National Research Council. 1999c. Making Climate Forecasts Matter. Panel on Human Dimensions of Seasonal-to-Interannual Climate Variability. National Academy Press, Washington, D.C.

National Research Council. 1999d. Sharing the Fish: Toward a National Policy on Individual Fishing Quotas. National Academy Press, Washington, D.C.

National Research Council. 1999e. Sustaining Marine Fisheries. Report NSB 99-133, Washington, D.C.

National Research Council. 1999f. Our Common Journey: A Transition Toward Sustainability. National Academy Press, Washington, D.C.

National Research Council. 1999g. The Use of Drugs in Food Animals: Benefits and Risks. Committee on Drug Use in Food Animals. National Academy Press, Washington, D.C.

National Research Council. 2000a. Clean Coastal Waters: Understanding and Reducing the Effects of Nutrient Pollution. National Academy Press, Washington, D.C.

National Research Council. 2000b. Seeing into the Earth: Noninvasive Characterization of the Shallow Subsurface for Environmental and Engineering Applications. National Academy Press, Washington, D.C.

National Research Council. 2001. Marine Protected Areas: Tools for Sustaining Ocean Ecosystems. National Academy Press, Washington, D.C.

National Science Board. 2000. Environmental Science and Engineering for the 21st Century: The Role of the National Science Foundation. National Science Foundation, Arlington, VA.

Netting, R., M. 1981. Balancing on an Alp. Cambridge University Press, Cambridge, U.K.

Norris, R. D., and U. Rohl. 1999. Carbon cycling and chronology of climate warming during the Palaeocene/Eocene transition. Nature 401:775-778.

Openshaw, S. 1983. The modifiable areal unit problem. Geo Books, Norwich, UK.

Organization for Economic Cooperation and Development. 1997. Sustainable Consumption and Production. ISBN 92-64-15515-5. Paris.

Ostrom, E. 1990. Governing the Commons: The Evolution of Institutions for Collective Action. Cambridge University Press, New York.

Ostrom, E., J. Burger, C. Field, R. Norgaard, and D. Policansky. 1999. Revisiting the commons: Local lessons, global challenges. Science 284:278-282.

Patz, J. A., W. J. M. Martens, D. A. Focks, and T. H. Jetten. 1998. Dengue fever epidemic potential as projected by general circulation models of global climate change. Environ. Health Prospect. 106(3):147-53.

Pielke, R. A., R. L. Walko, L. Steyaert, P. L. Vidale, G. E. Liston, and W. A. Lyons. 1999. The influence of anthropogenic landscape changes on weather in southern Florida. Monthly Weather Review 127:1663-1673.

Policansky, D. 1977. The winter of 1976-77 and the prediction of unlikely weather. Bulletin of the American Meteorological Association 58:1073-1074.

Policansky, D. 1999. Interdisciplinary problem-solving: The National Research Council. Policy Sciences 32:385-391.

Postel, S. L. 1999. Pillar of Sand: Can the Irrigation Miracle Last? W. W. Norton, New York.

Postel, S. L. 1998. Water for food production: Will there be enough in 2025? BioScience 48:629-637.

Preston, F. W. 1948. The commonness and rarity of species. Ecology 29:254-283.

Ramankutty, N., and J. Foley. 1999. Global Geochem. Cycles. 13:4.

Rhodes, L., C. Scholin, I. Garthwaite. 1998. *Pseudo-nitzschia* in New Zealand and the role of DNA probes and immunoassays in refining marine biotoxin monitoring programmes. Nat. Toxins 6(3-4):511.

Robbins, P. 1998. Authority and environment: Institutional landscapes in Rajasthan, India. Annals of the Association of American Geographers 88:410-345.

Roeder, P. L., W. N. Masiga, P. B. Rossiter, R. D. Paskin, and T. U. Obi. 1999. Dealing with animal disease emergencies in Africa: Prevention and preparedness. Rev. Sci. Tech. 18(1):59-65.

Rosenzweig, M. L. 1999. Species diversity. Advanced Theoretical Ecology: principles and applications, pp. 249-281. J. McGlade, ed. Blackwell Science, Oxford, UK.

Rosenzweig, M. L., and Y. Ziv. 1999. The echo pattern in species diversity: Pattern and process. Ecography 22:614-628.

Ross, P., R. De Swart, R. Addison, H. Van Loveren, J. Vos, and A. Osterhaus. 1996. Contaminant-induced immunotoxicity in harbour seals: wildlife at risk? Toxicology 112(2):157-69.

Rubin, E. S., L. B. Lave, and M. G. Morgan. 1992. Keeping climate research relevant. Issues in Science and Technology VIII(2):47-55.

Ruttan, L. M. 1998. Closing the commons: Cooperation for gain or restraint. Human Ecology 26:43-66.

Schimel D., I. G. Enting, M. Heimann, T. M. L. Wigley, D. Raynaud, D. Alves, and U. Siegenthaler. 1995. CO_2 and the Carbon Cycle. Pp. 35-71 in J. T. Houghton, L. G. Meria Filho, J. Bruce, H. Lee, B. A. Callander, E. Haites, N. Harris, and K. Maskel, eds. Climate Change 1994: Radiative Forcing of Climate Change and an Evaluation of the IPCC IS92 Emission Scenarios. Cambridge University Press, Cambridge U.K.

Schimel D., J. Melillo, H. Tian, A.D. McGuire, D. Kicklighter, T. Kittel, N. Rosenbloom, S. Running, P. Thornton, D. Ojima, W. Parton, R. Kelly, M. Sykes, R. Neilson, and B. Rizzo. 2000. Contribution of the increasing CO_2 and climate to carbon storage by ecosystems in the United States. Science 287:2004-2006.

Schlesinger, W. H. 1997. Biogeochemistry: An Analysis of Global Change, 2nd ed. Academic Press, San Diego.

Scholin, C. A., F. Gulland, G. J. Doucette, S. Benson, M. Busman, F. P. Chavez, J. Cordaro, R. DeLong, A. De Vogelaere, J. Harvey, M. Haulena, K. Lefebvre, T. Lipscomb, S. Loscutoff, L. J. Lowenstine, R. Marin III, P. E. Miller, W. A. McLellan, P. D. Moeller, C. L. Powell, T. Rowles, P. Silvagni, M. Silver, T. Spraker, V. Trainer, and F. M. Van Dolah. 2000. Mortality of sea lions along the central California coast linked to a toxic diatom bloom. Nature 403(6765):80-84.

Schuckert, M. 1997. Pp. 325-329 in Proc. 3rd Intl. Conf. on Ecomaterials. Society of Non-Traditional Technology, Tokyo.

Scott, M. R., R. Will, J. Ironside, H. O. Nguyen, P. Tremblay, S. J. DeArmond, and S. B. Prusiner. 1999. Compelling transgenic evidence for transmission of bovine spongiform encephalopathy prions to humans. Proceedings of the National Academy of Sciences 96(26):15137-15142.

Shimshony, A. 1999. Disease prevention and preparedness in cases of animal health emergencies in the Middle East. Rev. Sci. Tech. 18(1):66-75.

Silbergeld, E. K., L. Grattan, D. Oldach, and J. G. Morris. 2000a. *Pfiesteria*: Harmful algal blooms as indicators of human-ecosystem interactions. Environ. Res. 82(2):97-105.

Silbergeld, E. K., J. Sacci, and A.F. Azad. 2000b. Mercury exposure and murine response to *Plasmodium yoelli* infection and immunization. Immunopharmacology and Immunotoxicology 22:685-695.

Siy, R. Y., Jr. 1982. Community Resource Management: Lessons from the Zanjera. University of the Philippines Press, Quezon City.

Skole, D., and C. Tucker. 1993. Tropical deforestation and habitat fragmentation in the Amazon: satellite data from 1978 to 1988. Science 260:1905-1910.

Slovic, P. 1987. Perception of risk. Science 236:280-285.

Solley, W. B., R. K. Pierce, and H. A. Perlman. 1998. Estimated Use of Water in the United States, 1995. U.S. Geological Survey Circular 1200, Denver, CO.

Stavins, R. N. 1998. What can we learn from the Grand Policy Experiment?: Lessons from SO_2 allowance trading. Journal of Economic Perspectives 12(3):69-88.

Stein, B. A., and S. R. Flack. 1997. 1997 Species Report Card: The State of U.S. Plants and Animals. The Nature Conservancy, Arlington, VA.

Tans, P.P., I.Y. Fung, and T. Takahashi. 1990. Observational constraints on the global atmospheric CO_2 budget. Science 247:1431-1438.

Thomas, V., and T. Spiro. 1994. Emissions and exposure to metals: Cadmium and lead. In R. Socolow, C. Andrews, F. Berkhout, and V. Thomas, eds. Industrial Ecology and Global Change. Cambridge University Press, Cambridge, U.K.

Tilman, D. 1999. Ecological consequences of biodiversity: A search for general principles. Ecology 80:1455-1474.

Tolba, M. K., and O. A. El-Kholy, eds. 1992. The World Environment 1972-1992: Two Decades of Challenge. UNEP and Chapman & Hall, London.

Tollefson, L., S. F. Altekruse, and M. E. Potter. 1997. Therapeutic antibiotics in animal feeds and antibiotic resistance. Rev. Sci. Tech. 16(2):709-715.

Townsend, A. R., B. H. Braswell, E. A. Holland, and J. E. Penner. 1996. Spatial and temporal patterns in potential terrestrial carbon storage resulting from deposition of fossil fuel derived nitrogen. Ecol. Appl. 6(3):806-814.

Turner, M. G. 1990. Spatial and temporal patterns of landscape patterns. Landscape Ecology 4:21-30.

Turner, W., W. A. Leitner, and M. L. Rosenzweig. 2000. Ws2m.exe Web site http://turner.biosci.arizona.edu/ws2m/

Vesterby, M., A. Daugherty, R. Hemlich, and R. Claassen. 1997. Major Land Use Changes in the Contiguous 48 States. AREI UPDATES 1997. No. 3. USDA, ERS, NRED.

Vitousek, P. V., J. D. Aber, R. W. Howarth, G. E. Likens, P. A. Matson, D. W. Schindler, W. H. Schlesinger, and D. G. Tilman. 1997a. Human alteration of the global nitrogen cycle: sources and consequences. Ecological Applications 7(3):737-750.

Vitousek, P. M., H. A. Mooney, J. Lubchenco, and J. M. Melillo. 1997b. Human domination of Earth's ecosystems. Science 277:494-499.

von Humboldt, F. H. A. 1807 (1959). Essai sur la geographie des plantes. Sherbom Fund Facsimile 1. Society for the Bibliography of Natural History, London.

Wardle, D. A., M. A. Huston, J. P. Grime, F. Berendse, E. Garnier, W. K. Lauenroth, H. Setälä, and S. D. Wilson. 2000. Biodiversity and ecosystem function: An issue in ecology. Bulletin of the ESA. 81(3):235-239.

Wegener, H. C., F. M. Aarestrup, L. B. Jensen, A. M. Hammerum, and F. Bager. 1999. Use of antimicrobial growth promoters in food animals and *Enterococcus faecium* resistance to therapeutic antimicrobial drugs in Europe. Emerg. Infect. Dis. 5(3):329-335.

Wijkman, A. 1999. Sustainable development requires integrated approaches. Policy Sciences 32:345-350.

Williams, G. C., and R. M. Nesse. 1991. The dawn of Darwinian medicine. Q. Rev. Biol. 66(1):1-22.

Wilson, M. E. 1999. Emerging infections and disease emergence. Emerging Infectious Disease 5(2):308-309.

Young, O. R, with contributions from A. Agrawal, L. A. King, P. H. Sand, A. Underdal, and M. Wasson. 1999. Institutional Dimensions of Global Environmental Change: IDGEC Science Plan. International Human Dimensions Programme on Global Environmental Change, Bonn, Germany.

Zedler, J. 1996. Coastal mitigation in Southern California: The need for a regional restoration strategy. Ecological Applications 6:84-93.

Appendix A

Letter of Solicitation

**GRAND CHALLENGES IN ENVIRONMENTAL SCIENCES:
SEEKING INPUT**

FEBRUARY 1, 1999

A new committee of the United States National Research Council (Committee on Grand Challenges in Environmental Sciences) has been asked by the National Science Foundation to identify and prioritize grand challenges in environmental sciences that are likely to be relevant over the next 10-30 years. The committee would like advice from the scientific community and others. The idea is to describe a few grand challenges that appear to have the greatest scientific importance, research potential, and practical value. The definition of environmental sciences is broad, including the natural sciences, the social sciences, and engineering. A description of the project's scope and a list of committee members with brief biographical sketches will be available on the Web at <www.nas.edu/gces> as soon as this Web site is complete, about February 8, 1999.

The committee invites submission of ideas for grand challenges in environmental sciences; it will use those ideas to help it in its task, and anticipates inviting some respondents to a future workshop for more extended discussions. The time and place of the workshop will be announced on the committee's Web site.

The committee considers that any grand challenge must be compelling to scientists and the public, must require a sustained research effort, and must be intellectually exciting. Beyond that, we ask you to provide the following information about your candidate grand challenge:

A. A one-sentence summary that will quickly convey your idea to a broad audience.

B. A narrative description of the challenge.

C. Evaluation of the challenge in terms of all the following criteria that are relevant to your idea:

- Scientific payoff.
- Practical payoff (i.e., help in solving environmental problems).
- Feasibility (likelihood that valuable results would be produced over the next ten years).
- The need for interdisciplinary collaboration, especially if it builds capacity for dealing with other challenges.
- Research resources available or needed, including infrastructure such as new technology or information systems.
- The ability of existing institutions to support the research effort.

We encourage you to think broadly and to provide ideas that make connections among the sciences and between science and practical needs. The committee will begin to consider suggestions on February 26, 1999 and suggestions will be most useful if they are received by then. However, it is likely that suggestions received even as late as the beginning of May will be at least of some help to the committee. We ask for your suggestions, preferably by email to GCES@NAS.EDU. Please also provide your name, your affiliation (if any), your email address, your mailing address, and your telephone number.

We ask you to keep your suggestion to a maximum of one printed page. You may use the email address provided above for questions about our committee and our process as well as for your suggestions. You can also find information on the committee's activities and its report on its Web site.

We look forward to receiving your suggestions. We also encourage you to share this message with colleagues who may be interested.

> Committee on Grand Challenges in Environmental Sciences
> National Academy of Sciences/National Research Council
> Room HA-354
> 2101 Constitution Avenue, N.W.
> Washington DC 20418
> U.S.A.

Appendix B

Authors of Suggested Grand Challenges

Neil Adger, University of East Anglia

Harold Agnew, San Diego Supercomputer Center

Alwynelle Ahl, United States Department of Agriculture

William M. Alexander, California Polytechnic State University

Bernard Amos, Duke University

Joe Anderson, Paper, Allied-Industrial, Chemical, and Energy Workers International Union

Silvia Austerlic, Antioch University of Seattle

Paul Baker, Duke University

William A. Bartlett, Leonardtown, MD

Rebecca Belling, Virginia Polytechnic Institute and State University

Lee E. Benda, Earth Systems Institute

Heiner Benking, Forschungsinstitut Für Anwendungsorientierte Wissensverarbeitung Ulm, Germany

Helge Berglann, University of Oslo, Norway

R. Stephen Berry, University of Chicago

David L. Bish, Los Alamos National Laboratory

William Blomquist, Indiana University

Ken Bone, Del Mar High School, San Jose, CA

David Boyd, United States Fish and Wildlife Service

Thomas Bragg, University of Nebraska at Omaha

Kenneth H. Brink, Woods Hole Oceanographic Institute

David Brower

Gordon E. Brown, Stanford University

Anne Burrill, European Commission

Peter Buseck, Arizona State University

Fred Cagle, Pueblo of San Diego Watershed

Lynton K. Caldwell, Indiana University

Christopher Canaday

Valerie Chase, National Aquarium, Baltimore

H. H. Cheng, American Society of Agronomy

Stuart W. Churchill, University of Pennsylvania

Keith Clarke, University of California, Santa Barbara

John Cloud, University of California, Santa Barbara

Donna K. Cobb

Peter Collins, University of California, Santa Barbara

Cathy Copley, Illinois Department of Public Health

James A. Crutchfield, University of Washington (retired)

Claire Hope Cummings, Food and Farming Forum

Kevin P. Czajkowski, University of Toledo

Wim A. de Bruyn, ZERO, Brussels, Belgium

Joan E. Denton, California Environmental Protection Agency

Francis A. DiGiano, University of North Carolina at Chapel Hill

Denny Dobbin, North Carolina

William E. Easterling, Pennsylvania State University

Isidore S. Edelman, Columbia University

Paul R. Epstein, Harvard University Medical School

Gary W. Ernst, Stanford University

Ronald Estabrook, University of Texas

Rick Farnell, Yukon Department of Renewable Resources

John Felleman, State University of New York, Syracuse

Mark E. Fernau, Earth Tech, Inc., Concord, MA

Ann Fisher, The Pennsylvania State University

Gustavo Fonseca, Center for Applied Biodiversity Science, Conservation International

Ralph Franklin, Clemson University

Anne Frondorf, U.S. Geological Survey

Wilford Gardner, Utah State University

Daniel Gerber, University of Massachusetts

Lewis E. Gilbert, Columbia University

Robert H. Giles, Jr., Virginia Polytechnic Institute and State University

Ann Gill, Mecklenburg County, Charlotte, NC

Daniel Gilrein, Cornell Cooperative Extension of Suffolk County

Karl Glasener, AESOP Enterprises, Ltd.

W. Reid Goforth, U.S. Fish and Wildlife Service

Thomas Gold, Ithaca, NY

Marvin Goldberger, University of California, San Diego

Dan Golomb, University of Massachusetts, Lowell

William Gordon, Rice University

Carmine Gorga

Charles Greene, Cornell University

Robert J. Gregory, Massey University, Palmerston North, New Zealand

Jerry Gross, Harvard University School of Medicine

Gordon Haas, British Columbia Ministry of Fisheries

Peter M. Haas, University of Massachusetts

William Harmon, Carrollton, TX

James Harsh, Washington State University

Richard Hegg, United States Department of Agriculture-Cooperative State Research, Education, and Extension Service

Grant Heiken, Los Alamos National Laboratory

William D. Henriques, Agency for Toxic Substances and Disease Registry

Colin Henry, Deakin University, Geelong, Victoria, Australia

Robert A. Herendeen, Illinois Natural History Survey

William H. Herke, Louisiana State University

David Hicks, University of Bath

Frank Hole, Yale University

James Holton, University of Washington

Ted Howard

John Humphreys

Daniel H. Janzen, University of Pennsylvania

Thomas C. Johnson, University of Minnesota

Kenneth Y. Kaneshiro, University of Hawaii

John Kappenman, Minnesota Power Company

Mick Kelly, University of East Anglia

Karen Kempster, Rancho Cordova, CA

Robert W. Keyes, IBM Thomas J. Watson Research Center (Emeritus)

Roland A. Knapp, University of California

Walter Kohn, University of California, Santa Barbara

Doug La Follette, State of Wisconsin

Sam Lammie, U.S. National Park Service

Louis J. Lanzerotti, Lucent Technologies

Henry Lardy, University of Wisconsin

Alexander Leaf, Harvard University

W. Christopher Lenhardt, Columbia University

Mike Levine, University of California, Berkeley

Richard Levins, Harvard University

William Lewis, University of Colorado

Y. Q. Li

R. K. Liu

Max Loewenstein, National Aeronautics and Space Administration

Robert Loewy, Georgia Tech University

David E. Loper, Florida State University

Nick Loux, U.S. Environmental Protection Agency

Richard Lowenberg, Davis Community Network & Yolo Area Regional Network

Mark Lubell, State University of New York, Stony Brook

Harvey F. Ludwig, Seatec International Consulting Engineers, Bangkok, Thailand

Roger Lukas, University of Hawaii

Richard Luthy, Stanford University

Michael C. MacCracken, United States Global Change Research Program

Luisa Maffi, Northwestern University

Thomas F. Malone, Connecticut Academy of Sciences and Engineering

Greg Mandt, National Oceanic and Atmospheric Association

Carl N. McDaniel, Rensselaer Polytechnic Institute

Michael Vincent McGinnis, University of California, Santa Barbara

Alan McIntosh, University of Vermont

Peter Meisen, Global Energy Network Institute

William Merrill, Smithsonian Institution

Dwight F. Metzler, Topeka, KS

Daniel J. Miller, Earth Systems Institute

Roberta Balstad Miller, Columbia University

Hugh H. Mills, Tennessee Technological University

James I. Mills, Lockhead-Martin Idaho Technologies

Forrest M. Mims III, Sun Photometer Atmospheric Network

François Morel, Princeton University

Lowell E. Moser, Crop Science Society of America

Richard A. Muller, University of California, Berkeley

Arthur Noll, North Vassalboro, ME

Charles O'Melia, The Johns Hopkins University

Michael O'Neill, U.S. Department of Agriculture

Lenora A. Oftedahl, Columbia River Inter-Tribal Fish Commission, Portland

George A. Olah, University of Southern California

Antoni K. Oppenheim, University of California, Berkeley (Emeritus)

Stan A. Orchard, Canadian Amphibian and Reptile Conservation Network

Elisabeth Pate-Cornell, Stanford University

Anthony Patt, Harvard University

William George Paul, Virginia Polytechnic Institute and State University

Merrilynn J. Penner, University of Maryland

Naraine Persaud, Virginia Polytechnic Institute and State University

S. George Philander, Princeton University

Ronald L. Phillips, Crop Science Society of America

John A. Piatt, Pacific Northwest National Laboratory

Incigul Polat, Scientific and Technical Research Council of Turkey

Franklin Wayne Poley, Culturex, Vancouver, BC

Warren Porter, University of California, Santa Barbara

Darrell Posey, Oxford University

John Pozzi, Global Resource Bank

Robert A. Rapp, Ohio State University

David J. Rapport, International Society for Ecosystem Health

Christopher J. Raxworthy, University of Kansas

Peter Rhines, University of Washington

Ed Rhodes, National Oceanic and Atmospheric Association

Robert Rhodes, Baton Rouge, LA

Clifford P. Rice, United States Department of Agriculture, Agricultural Research Service

Elizabeth Rieben, Bureau of Land Management

Norman Robbins, Case Western Reserve University School of Medicine

Eugene A. Rosa, Washington State University

Norbert Ross, Northwestern University

Daniel Sarewitz, Columbia University

A. Cemal Saydam, Middle East Technical University

Raymond Schmitt, Woods Hole Oceanographic Institution

Stephen H. Schneider, Stanford University

Martin E.P. Seligman, University of Pennsylvania

Hamide Senyuva, Scientific and Technical Research Council of Turkey

Robert Serafin, National Center for Atmospheric Research

Julia Shaw, The Swire Institute of Marine Science, Hong Kong

S. Fred Singer, George Mason University

Fred Sissine, Congressional Research Service

Ian Small, Medicins Sans Frontieres, Tashkent, Uzbekistan

Dale R. Smith, University of Georgia

Margaret E. Smith, Cornell University

Lee Snyder, U.S. Air Force Consultant

Susan Solomon, National Oceanic and Atmospheric Association

Richard E. Spalding, Sandia National Laboratories

Suzanne Spradling, Oklahoma State University

Franklin W. Stahl, University of Oregon

Chauncey Starr, Electric Power Research Institute

Theodore L. Steck, University of Chicago

Richard S. Stein, University of Massachusetts

Jeffrey I. Steinfeld, Massachusetts Institute of Technology

Lawrence E. Stevens, United States Geological Survey

Richard Stock, University of Dayton

Steven H. Strauss, Oregon State University

Kristi L. Sullivan, Cornell University

Art Sussman, WestEd, San Francisco

Stephen R. Sutton, University of Chicago

Johanna Krout Tabin

Mark Thiemens, University of California, San Diego

Chris Tirpak, U.S. Environmental Protection Agency

Chadwick A. Tolman, National Science Foundation

Gail Tonnesen, University of California, Riverside

Cheryl Travis, University of Tennessee, Knoxville

Kevin E. Trenberth, National Center for Atmospheric Research

R. Rhodes Trussell, Montgomery Watson, Inc., Pasadena, CA

Roger Y. Tsien, University of California, San Diego

Greg A. Valentine, Los Alamos National Laboratory
Joost van der Meer, Medecins Sans Frontieres
James R. Van Sant, Oakdale, CA
Lisa M. Vandemark, Rutgers University
Pier Vellinga, Vrije Universiteit, The Netherlands
Stuart Wagenius, University of Minnesota
Jeff Waldon, Virginia Polytechnic Institute and State University
John M. Wallace, University of Washington
Jaw-Kai Wang, University of Hawaii at Manoa
Bess Ward, Princeton University

Mary E. Watson, North Carolina Geological Survey
Wilford F. Weeks, University of Alaska (retired)
Carol Werner, University of Utah
Andre Wille, Colorado Division of Wildlife
Carl Woese, University of Illinois
George M. Woodwell, The Woods Hole Research Center
Chang-Yu Wu, University of Florida, Gainesville
Peter J. Wyllie, California Institute of Technology
Dan Yurman, Idaho Falls, ID
Mary Lou Zoback, United States Geological Survey